THE ART OF WAR
Organized for Decision Making

By Brace E. Barber
3rd Edition

Information about the **Executive Create the Know Program** and Brace E. Barber's availability for speaking and training can be found at www.NoExcuseLeadership.com

Copyright © 2016 - 2018 Patrol Leader Press and Brace E. Barber
ISBN 0-9678292-5-9

All Rights Reserved. No portion of this book may be reproduced, stored in a retrieval system, or transmitted in any form or by any means - electronic, mechanical, photocopy, recording, scanning, or other - except for brief quotations in critical reviews or articles, without the prior written permission of the publisher.

®Patrol Leader Press, Thompson's Station, TN 37179. b@mkoi.us

Unless otherwise indicated, *The Art of War* book and references are from the 1910 translation by Lionel Giles, M.A.

Other works by Brace E. Barber
- *TESTED: A New Strategy for Keeping Kids in the Faith.* (Patrol Leader Press, May 2018) www.BraceBarber.com
- *No Excuse Leadership; Lessons from the U.S. Army's Elite Rangers* (J. Wiley and Sons 2003) www.NoExcuseLeadership.com
- *Ranger School, No Excuse Leadership* (Patrol Leader Press, Inc. 2000) www.RangerSchool.com

Ongoing resources and information can be found at
www.NoExcuseLeadership.com

Patrol Leader Press, Inc.
"Move out and draw fire."

With regard to the complete The Art of War, included at the end of this book, translated from the Chinese By Lionel Giles, M.A. (1910) and reprinted through The Project Gutenberg, ONLY this portion of the book is for the use of anyone anywhere at no cost and with almost no restrictions whatsoever. You may copy it, give it away or re-use it under the terms of the Project Gutenberg License or online at www.gutenberg.org
Release Date: December 28, 2005 [eBook #17405]
Language: English
Character set encoding: ISO-8859-1

Dedication

This book is dedicated to my wife, Natasha. She seeks God first and maintains her strength through His grace and the love of Jesus Christ. It is only because of her love, support and encouragement that I can endure the wild ride of entrepreneurship and creation.

Contents

Foreword 6
Acknowledgements 9
About the Author 10
Chapter 1: ADVANCED STRATEGIST 11
Chapter 2: SUPREME EXCELLENCE 22
Chapter 3: FIVE FACTORS, 10 QUESTIONS 24
Chapter 4: DECISION MAKING FOUNDATION 27
Chapter 5: FRIENDLY MISSION 31
Chapter 6: TERRAIN and GROUND (Earth) 33
Chapter 7: ENEMY 52
Chapter 8: STRENGTH 60
Chapter 9: POSSIBLE ACTIONS 64
Chapter 10: HEAVEN 75
Chapter 11: ATTACK BY FIRE 77
Chapter 12: DECEPTION PLAN 80
Chapter 13: FULL SPECTRUM DECISION MAKING 84
CONCLUSION to *The Pattern of Power* 92
THE ART OF WAR 94
I. LAYING PLANS 94
II. WAGING WAR 96
III. ATTACK BY STRATAGEM 98
IV. TACTICAL DISPOSITIONS 100
V. ENERGY 102
VI. WEAK POINTS AND STRONG 104
VII. MANEUVERING 107
VIII. VARIATION IN TACTICS 110
IX. THE ARMY ON THE MARCH 112
X. TERRAIN 116
XI. THE NINE SITUATIONS 119
XII. THE ATTACK BY FIRE 125
XIII. THE USE OF SPIES 126

Foreword

By Art Jacobs

During a psychology class in school a long time ago, my professor told us that humans cannot survive a single day without rationalizing – the mental exercise of explaining (or justifying to ourselves) our behavior and the things we think, ponder, encounter, deal with, or have to confront and overcome.

And equally, I have come to realize that humans cannot survive a single day without planning. And, implicit in planning, and integral component to planning is the act of strategizing. We have "macro" strategies for our lives (educational, career, financial, and marital to name just a few examples). But we also have everyday "micro" strategies for our lives (what we will tackle at work that day and how, organizing birthday parties and holidays, navigating the aisles at the grocery store, and even when and how we will cross a busy street to cite a few examples.

Many of the everyday strategies are second nature, repetitive, involve muscle-memory, and rather simple. We do not devote much time (nor should we) to many of them. That's okay.

Plans and strategies surround us all – either ones we have formulated and implement ourselves, the ones we willingly implement for others (as in our job), and the ones must implement or comply with because we must (doing homework in school, making mortgage payments, filing and paying taxes).

Most people feel empowered because they are in reasonable control of their plans and strategies. Many others however, feel rather powerless, trapped, or even victimized by the plans and strategies of others.

Given that planning and strategizing are critical to our very survival, success, fulfillment, and well-being, why is it that a majority of people do not study planning and strategizing as a science or skill? The same people who seem to complain the most about the plans they must live with, seem to be the ones least likely to acquire the knowledge and ability to plan better.

If you are among the group of people who believe that planning and strategizing is important, this is probably not the first book on the subject you have read. For those people (or even the ones reading their first tome on the topic) who are intent on embarking on a more definitive study of the topic called strategy, you hold in your hand a uniquely excellent source.

So, why Sun Tzu? Allow me a simple argument for the case: History is filled with examples of where smaller armies have defeated larger armies. It's never been about the weapons and number of soldiers, but about the strategies and morale. Just because you invent a product does not mean that you will dominate a market (IBM did not invent the computer, Intel did not invent the integrated circuit, Visa did not invent the credit card, Microsoft did not invent software, and Wal-Mart did not invent big stores).

Many companies have been founded, were initially very successful, were doubling in size each year, and were even the darling of Wall Street. And now they are bankrupt, because of either poor planning or strategizing, or someone else, with a sometimes-lesser product, had a superior plan or strategy.

My point is that when you really think about it – there have never really been any "strategic products" – only "strategic thinking." So, if you trace the history of "strategic thinking" you eventually end up 2,500 years ago in China with Sun Tzu (philosopher, advisor, court counsel, priest, and general).

Sun Tzu said, "The key to victory is not in defeating the enemy, but in defeating the enemy's strategy; therein lies not only their vulnerability, but the sublime essence of strategy." What he is saying that it is the tactician who would attack a person, place, product, or company. Those "so-called plans" are usually obvious, often resource-intensive, easily protracted, and relatively straightforward to defend against.

Hopefully, you have found what's been covered so far to be at least intriguing. I can assure you that if you are serious about becoming more proficient at being strategic, studying Sun Tzu is always both the best first step, and a continuing quest.

Brace Barber has ingeniously combined his military experience, history, the incredible insights of Sun Tzu, and the challenges the modern business executive today must contend with in this chaotic and ever-changing world. Anyone wishing to formulate a more pragmatic means to making important decisions, or who would like to think and act "more strategically" should arm themselves with this book and the principles contained herein.

Art Jacobs is the founder and CEO of the Valkyrie Consulting Group, LLC, a highly renowned and successful international sales and sales management training and consulting firm. Art Jacobs was also a highly decorated Captain in the United States Army, was the Aviation Detachment Commander for a Special Forces (Green Beret) Group, served two combat tours flying helicopters in Vietnam, was a NATO Intelligence and Operations Officer.

Acknowledgements

For ALL things, including the strength to persevere with peace, I thank God.

I am thankful to Art Jacobs for reintroducing *The Art of War* to me in 2005. He gave me the book at a point where the teachings were able to take root in the tilled experiences of my life.

I am thankful to Felix Carabello, who at approximately the same time became a wonderfully motivating thought partner in the exploration of complex decision-making and strategy.

About the Author

Brace E. Barber was most recently the CEO of General Stability, Inc. a holding company operating in the tactical gear and security market. He was the co-founder and CEO of M9 Defense, Inc., a material technology company. He is formerly the President/CEO of Decipherst, Inc. and President of Tax Recovery Group, Inc. He has worked extensively in the field of multi-criteria decision making, strategy and leadership development, with a focus on individual performance while under extreme duress. He is the author of the books, *No Excuse Leadership, Lessons from the U. S. Army's Elite Rangers* and *TESTED: A New Strategy for Keeping Kids in the Faith*. He performs limited executive coaching through his *Create the Know* program, which takes executives through the analysis and course of action development process to solve overwhelming problems. Brace is a West Point graduate and veteran of 11 years of active military service. He has earned military awards including the coveted ranger tab and airborne wings. He lives in middle Tennessee with his wife and three children.

Chapter 1: ADVANCED STRATEGIST
See What Others Do Not See

Previously Unpublished excerpt:
"Master Tzu sat at a cluttered table in a stark room in the home of the Administrator of the Ch'i state. He bowed slightly over his work in what must have been last minute contemplation of the calculations he had made for his army's movement the next morning. From the entryway, I watched. I sensed vulnerability in his demeanor, which was completely uncommon except just before he sends men to their death in battle.

When he stood, he took all of his designs and left the room without looking at me. A small roll of parchment, bound with a strap, remained on the edge of the table. I knew what he had left and my duties with its regard. I knew the work to which he had dedicated himself over the past year and how concerned he was over its release.

He was pressed by the lives of the men in the barracks and a sovereign's deadline.

It was to the King of Wu that I delivered the package."

See the Entire Philosophy as a Whole for the First Time
This organization of The Art of War is powerful and can be used to achieve results, whether you are seeking organizational or personal goals. By releasing this work to you I am giving you the ability to have Sun Tzu himself counsel you in your actions. Please be careful how you use it.

The biggest risk is that you will misinterpret some of the teachings and put yourself into a compromising situation.

Specifically, some will interpret the statement that all war is based on deception (LP 18), as a license to lie and cheat or induce others to do so. Some may see it as a validation of their current habits of deception. If you use this material in this way you will not succeed and may even cause damage from which your bank account or reputation are unable to recover - or worse.

With this in mind, I caution you to take heed to the first of the seven considerations by which Sun Tzu can predict victory or defeat. He asks which of the two sovereigns is imbued with the Moral Law. (LP 13) Dismiss this key predictor at your own peril.

Does this statement taken into account with the previous, all war is based on deception (LP 18); presume that the ends justify the means? That is your call.

I am convinced that you can avoid the pitfalls of misuse of this material by studying it as a whole. See the philosophy the way it was intended, where each part acts upon the other instead of the popular habit of seeing it as a set of slick daily quotes. The ability to quote *The Art of War* without an equal ability to fully understand context, is like shooting a revolver loaded with one round. Most of the time it is useless, and the one time it does fire, it kills you. A major problem with *The Art of War* as it exists in all previous English translations and business adaptations is that it is nearly impossible to see the completeness of the philosophy. The treatment given to the writing for the past 100 years has primarily been academic, with a recent sprinkling of the business-flippant.

Neither handling of the material allows you to consider the whole work in your daily pursuits. The work you hold in your hands changes everything. This *Art of War*, for the first time in Western Society, organizes the classic work into a functional, usable tool for you.

You are the General. You are the warrior leader, and I will treat you that way, so despite my primary concerns, I release

The Art of War, Organized for Decision Making to you. But first, heed some supporting guidance from Sun Tzu himself. Use it wisely.

There are five dangerous faults which may affect a general: (1) recklessness, which leads to destruction; (2) cowardice, which leads to capture; (3) a hasty temper, which can be provoked by insults; (4) a delicacy of honor which is sensitive to shame; (5) over-solicitude for his men, which exposes him to worry and trouble. (VIT 12) These are the five besetting sins of a general, ruinous to the conduct of war. (VIT 13) When an army is overthrown and its leader slain, the cause will surely be found among these five dangerous faults. Let them be a subject of meditation. (VIT 14)

The consummate leader cultivates the moral law, and strictly adheres to method and discipline; thus it is in his power to control success. (TD 16) The general who advances without coveting fame and retreats without fearing disgrace, whose only thought is to protect his country and do good service for his sovereign, is the jewel of the kingdom. (T 24) No ruler should put troops into the field merely to gratify his own spleen; no general should fight a battle simply out of pique. (ABF 18) The general, unable to control his irritation, will launch his men to the assault like swarming ants, with the result that one-third of his men are slain, while the town still remains untaken. Such are the disastrous effects of a siege. (ABS 5)

He wins his battles by making no mistakes. Making no mistakes is what establishes the certainty of victory, for it means conquering an enemy that is already defeated. (TD 13) Thus it is that in war the victorious strategist only seeks battle after the victory has been won, whereas he who is destined to defeat first fights and afterwards looks for victory. (TD 15)

We are not fit to lead an army on the march unless we are familiar with the face of the country - its mountains and forests, its pitfalls and precipices, its marshes and swamps. (M 13)

The Reorganization
Is Sun Tzu's *The Art of War* incomplete? To the western mind it might seem so. The chapters are somewhat disjointed, many suggestions are repetitive and it is difficult to determine an order of priority for the philosophy.

To say that the work was incomplete or at least unorganized would be a defendable argument. For our purposes, however, I rest on its 2,500 years of successful application by kings, generals, soldiers and politicians as evidence that the philosophy is, in fact, complete. However, I will not sidestep the fact that the existing translations can be frustrating for the non-scholar to proactively use. I found the solution to understanding the completeness of the philosophy, and therefore its practical application, rests not in the translation, but in the organization. Through this new organization, you are able to understand the completeness and from that understanding you will see the pattern of power, which you can apply to your life's purposes.

Sun Tzu wrote a brilliant Taoist-based, Chinese text for a society that understood Taoism and war 2,500 years ago. Not only did the readers not have to negotiate the hazards of translating the language, they did not have to question the connections between the principles. They understood how everything fit together. In that time, *The Art of War* **was** organized.

To use a metaphor, Sun Tzu created the directions for a chariot, and everyone who read it in his culture was able to follow the directions to build a chariot. Our challenge today is that we do not need a chariot, and even if we did, we would have to outsource to China to build it. What we need is a bicycle. The same two wheels and ability to move, but something we can actually put on the 21st century road. The text is brilliant enough for that also, but it is not in the translation, it is in the organization of the directions. I have taken the two wheels and aligned them. I have taken the frame and elongated it. Without changing a word or diluting its effectiveness I have found the sprockets and gears and handles and peddles.

The purpose of this book is to allow you to actually use the process in your daily pursuits.

Since it is a bike and not a chariot, there will be some unneeded pieces left over; horsewhips, standing platform, gilded armor and helmet. For example, Sun Tzu said, When you come to a hill or a bank, occupy the sunny side, with the slope on your right rear. (AOM 13) Anyone claiming to know how to make that suggestion relevant today has a better imagination than I do. You will find my interpretations grounded and far from esoteric.

Interestingly, the purpose of this book is not to allow you a deeper understanding of *The Art of War*, or provide a new academic nuance to Sun Tzu's classic work. That has been done too many times by people qualified to do that. The purpose of this book is to allow you to actually use the process in your daily pursuits. Ironically, through the proper and repeated application of the work, you will in turn have a deeper understanding and appreciation of its nuances.

By using this first-of-a-kind organization you will be able to use Sun Tzu's own thought processes to develop enlightened and powerful courses of action (COA) that Sun Tzu himself would have suggested for the accomplishment of your goal. You will see the connection between different parts of *The Art of War*. You will know what priority conditions you need to meet, how they relate to each other, and be able to link those with Sun Tzu's own suggestions. You will understand the relevancy of *The Art of War* to today. You will accomplish your goals easier and you will see the patterns of power on the battlefield, in politics, in the office and in international relations. The brilliance of some leaders and the ignorance of others will come to light in astonishing color. Making decisions and plans using Sun Tzu's own process is now possible with very little uncertainty. I refer to this reorganization at the *Pattern of Power*.

> Competition is a reality. You can ignore that and compete at a disadvantage or learn how to create victory, which is available through *The Art of War*. Victory will ultimately elude those who are unwilling to contemplate and act in accordance with the deeper teaching of Sun Tzu and they will be crushed by the incompleteness of their plans and ignorance or ignoring of the moral law.

Reorganization Framework

We must understand that the purpose of *The Art of War* is to accurately guide people in their planning and decision making. The product of the application of the philosophy will be a thoroughly analyzed and constructed plan that is in concert with the natural laws of success. That is the purpose of the this organization as well.

I have reorganized *The Art of War* into four major areas; Planning process, Priorities, Expansion and Analysis, and Decision Making.

1. Planning Process: Our first step is to address the overarching planning process within which *The Art of War* fits. This planning process is the framework for organizing the entire philosophy into a useful tool and it encompasses the next

three areas of priorities, expansion and analysis and decision making.
2. Priorities: We will identify and look at the most important tenants of the philosophy. There are definite priorities expressed by Sun Tzu that we each must consider when determining the best COA. You will see these embodied in the chapters discussing Supreme Excellence and the Five Factors and Ten Questions.
3. Expansion and Analysis: Several chapters will expand on each of the priority areas with a compilation of related suggestions from *The Art of War* as expressed in Sun Tzu's own words.
4. Decision Making: Finally, we will examine a method for deciding from among the options, based on our situation and what the philosophy says is possible.

From the very beginning of our plunge into this work, you will gain insights that will advance your strategic planning competency. For the balance of this chapter we will look at the first part of our reorganization, the Planning Process.

Planning Process:
As part of our first step, Planning Process, we will learn how *The Art of War* fits into a larger decision making process. You will also see how the other three areas of our reorganization framework are included under this heading.

In relation to planning, Sun Tzu says;
1. Supreme excellence consists in breaking the enemy's resistance without fighting. (ABS 2)
2. The general who wins a battle makes many calculations in his temple ere the battle is fought. It is by attention to this point that I can foresee who is likely to win or lose. (LP 26)

3. If you know the enemy and know yourself, your victory will not stand in doubt; if you know Heaven and know Earth, you may make your victory complete. (T 31)
4. All war is based on deception (LP 18)

I will summarize these key strategy underpinnings in the following way.
- Mission and Purpose (Priorities)
- Planning and Foreknowledge (Expansion and Analysis)
- Deception

You will only encounter a handful of people in your life who truly grasp and do the hard work necessary to fully build out these four areas. Everyone else bounces around in the dark world of activity, rarely getting it right.

By reading the next few paragraphs, you will take a quantum leap in your strategic IQ. You will leave the hit-or-miss process of success behind and join the rare group that occupies the high ground looking down on the masses.

Once you receive a directive from your boss, the first thing you must do as an advanced strategist is write down the answer to the question, "What do I want to accomplish when all is said and done." That becomes your mission. Your mission does not have to be some grandiose life-long or world-changing goal, though it can be. Then by answering the question, "Why must I accomplish this mission," you are defining your purpose. It is critical for you to understand that in everything you do, from the next phone call to a prospect to the setting out of corporate goals, you must have a mission and a purpose. The answers to these questions will guide all of your planning.

Sun Tzu adds difficulty to the concept of mission when he tells us that supreme excellence consists in breaking the enemy's resistance without fighting. (ABS 2) He is setting an incredibly high standard for how we accomplish our mission and fulfill our purpose. Today, we would rather gear up and meet our competition in mortal combat. There is a better way, but it must be set as a purpose, with respect for the moral law that says, to shatter and destroy it is not so good. (ABS 1) Once our mission and purpose are set, **we must engage in the brutal, tiring and hard work of planning**.

Proper planning can be an organized and ordered process even for multi-dimensional and complex problems. If you do not know the process, you will make the common mistake of jumping right into creating a solution. Most are unaware that this mistake is a drastic shortcut which stifles innovation and betrays us the best solution.

The biggest mistake made by professionals is that they perform cursory planning overloaded with their assumptions and intuition and jump right to course of action development. When I work with clients, I watch as they struggle with this habit of jumping ahead, but I also see their eyes light up when valuable truth is revealed through the properly ordered format. The practical application of the process does everything a business owner or executive wants to produce; faster progress, reduced costs, fewer resources, more profit and all with fewer headaches.

Take hold of the fact that there are many steps of thought and analysis before you begin developing a solution, and you are advancing rapidly as a strategist. It is not necessary to know the steps just yet, simply be aware that they exist.

Much of the planning process revolves around foreknowledge; around the gathering of information - lots of information. Lots more information and in greater detail than you can imagine. Accurate and timely knowledge about all of the aspects of the battlefield and enemy are necessary for you to be able to achieve a supremely excellent victory. It is this information that you will analyze. And again, you will be in exclusive company if you do this properly.

Finally, deception. I include this here because it ranks at the top as a victim of neglect in most planning processes. Even the U.S. Army, which is a world-class planning organization, often treats deception as an afterthought. In pursuit of achieving your victory, your master plan is nobody else's business. Right?! Most people naturally attempt secrecy and think they are done with deception. But, Sun Tzu says, hence, when able to attack, we must seem unable; when using our forces, we must seem inactive; when we are near, we must make the enemy believe we are far away; when far away, we must make him believe we are near. (LP 19) He uses the words, "must make." He did not say we "must hope," he says "must make." Deception is both active, and passive. Get that point and then make sure you include active deception as part of your planning.

The operations of the U.S. Army have been deeply influenced by the teachings of Sun Tzu. It is not surprising then that the U.S. Army is exceptional at fulfilling these four pillars of strategy. They have developed a planning method called the multi-criteria decision making process (MCDMP) that allows the leadership team of an organization to effectively function, even under extreme pressure and information overload. This is important to us because it provides us with a proven model to follow that expands the practicality of *The Art of War* philosophy.

This model is so effective that I would venture to say it is unmatched, not even approached, in corporate America. Even after two weeks of 18-hour training days at the National Training Center (NTC), the Battalion staff of an Army unit would impress any CEO in the country with their planning success. I would go further to say that that tired and worn leadership team would relatively out perform nearly any team of well-rested, well-fed, and fully-caffeinated corporate team. Part of my confidence rests in the model, but the other part rests in the Army's unending dedication to, and the practice of, the model. The MCDMP is not blow-your-mind innovative but it will be here tomorrow and it will be the standard, and everyone knows it.

I have introduced this to you now so you can see the strategy pillars of *The Art of War*, that they are fully applicable to today, and that the MCDMP is a valid model for you to use to fulfill these areas. Recognize that *The Art of War* is a subset of the MCDMP.

> The *Pattern of Power,* tied to the decision making process, allows you to complete the process of taking *The Art of War* and using it properly in your life.

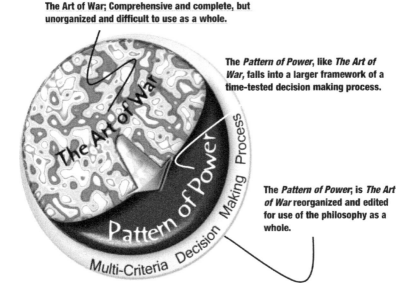

The Art of War; Comprehensive and complete, but unorganized and difficult to use as a whole.

The *Pattern of Power*, like *The Art of War*, falls into a larger framework of a time-tested decision making process.

The *Pattern of Power,* is *The Art of War* reorganized and edited for use of the philosophy as a whole.

Chapter 2: SUPREME EXCELLENCE
Priorities; Part 1

Sun Tzu uses various techniques to order the importance of his statements. One way is through a statement's direct correlation to victory or defeat. Another is through the use of adjectives such as *supreme*. It should be significant to you that I am starting with his highest description, Supreme Excellence.

Let this emphasis color the remainder of your study.

- Supreme excellence consists in breaking the enemy's resistance without fighting. (ABS 2)
- In war, then, let your great object be victory, not lengthy campaigns. (OWW 19)
- In the practical art of war, the best thing of all is to take the enemy's country whole and intact; to shatter and destroy it is not so good. (ABS 1)
- But a kingdom that has once been destroyed can never come again into being; nor can the dead ever be brought back to life. (ABF 21)

... So, too, it is better to recapture an army entire than to destroy it, to capture a regiment, a detachment or a company entire than to destroy them. (ABS 1) It is only one who is thoroughly acquainted with the evils of war that can thoroughly understand the profitable way of carrying it on. (WW 7) Therefore the skillful leader subdues the enemy's troops without any fighting; he captures their cities without laying siege to them; he overthrows their kingdom without lengthy operations in the field. (ABS 6) With his forces intact he will dispute the mastery of the Empire, and thus, without losing a man, his triumph will be complete. This is the method of attacking by stratagem. (ABS 7)

Thus the highest form of generalship is to
1. balk the enemy's plans;
2. the next best is to prevent the junction of the enemy's forces;
3. the next in order is to attack the enemy's army in the field;
4. and the worst policy of all is to besiege walled cities. (ABS 3)

Attacking the enemy is the lowest form of generalship, yet it is in these categories where companies spend most of their planning and operational energy. It is macho to be a corporate war fighter and a conqueror, but it is not through those approaches that you build shareholder value. Victory through leverage created by careful, intentional and thorough planning allows returns far greater with respect to the invested resources. Our strategists should be dedicated to defeating the enemy by the first two methods, and it is through learning the pattern of power that you are becoming just such an advanced strategist.

Our first priority is to achieve a supremely excellent victory.

Chapter 3: FIVE FACTORS, 10 QUESTIONS
Priorities; Part 2

These five factors and seven of the ten following questions are presented at the beginning of *The Art of War*. They are a good start in our reorganization because they are Sun Tzu's own priority list and they give us a central point around which to build. These are the measuring sticks by which we can tell if a general will achieve a supremely excellent victory.

The Art of War is governed by five constant factors, to be taken into account in one's deliberations, when seeking to determine the conditions obtaining in the field. (LP 3)

1. **The Moral Law** Causes the people to be in complete accord with their ruler, so that they will follow him regardless of their lives, undismayed by any danger. (LP 4-6)
2. **Heaven** Night, Day, Cold, Heat, Times, Seasons (LP 4, 7)
3. **Earth** Great Distances, Small Distances, Danger, Security, Open Ground, Narrow Passes, Chances of Life, Chances of Death (LP 4, 8)
4. **The Commander** Wisdom, Sincerity, Benevolence, Courage, Strictness (LP 4, 9)
5. **Method and Discipline** the marshalling of the army in its proper subdivisions the graduations of rank among the officers the maintenance of roads by which supplies may reach the army. (LP 4, 10)

By means of these seven considerations I can forecast victory or defeat. (LP 14)
1. Which of the two sovereigns is imbued with the Moral law? (LP 13)
2. Which of the two generals has most ability? (LP 13)
3. With whom lie the advantages derived from Heaven and Earth? (LP 13)
4. On which side is discipline most rigorously enforced? (LP 13)
5. Which army is stronger? (LP 13)
6. On which side are officers and men more highly trained? (LP 13)
7. In which army is there the greater constancy both in reward and punishment? (LP 13)

The following three considerations are included with the original seven because of Sun Tzu's own relating of their importance. Specifically, they are each highlighted as necessary to forecasting victory or defeat.

8. **Which general makes the most calculations before the battle is fought?** The general who wins a battle makes many calculations in his temple ere the battle is fought. It is by attention to this point that I can foresee who is likely to win or lose. (LP 26)
9. **Which sovereign refrains from interfering with his general?** Thus we may know that there are five essentials for victory" ...(5) He will win who has military capacity and is not interfered with by the sovereign." (ABS 17) The sovereign's *interference* is the only of the five essentials not covered in the other nine considerations. Therefore, if it is an essential for victory, along with the other characteristics, by determining if there is interference or not, we should be able to predict victory or defeat.

10. Which general has greater foreknowledge? Thus, what enables the wise sovereign and the good general to strike and conquer, and achieve things beyond the reach of ordinary men, is foreknowledge. (UOS 5) **Without foreknowledge you will be unable to strike and conquer.**

Using Sun Tzu's own detailed decision-making principles, which constitute the remainder of the book, we will answer the questions posed in 1, 3, 5, and 10. And though the development of a COA does not precisely cover the other questions, you will realize that through the application of this process, not only will you gain the necessary advantages, but you will more than satisfy the other considerations. By completing the planning process you are also fulfilling number 8 and proving number 2. These are the planning and strategy areas of the process and are mostly in the control of the leader.

The remaining considerations, 4, 6, 7 and 9, relate to the capabilities of the maneuver forces and though they are critical to victory, they are not the focus of our work. Our purpose is to use the pattern of power to determine when, where, and how to maneuver our forces. If you are attacking up hill against a superior force, these leadership factors are not important. How to achieve supreme victory is our focus.

In today's talent-driven world, it has been argued that all teams are equal, it is how you maneuver which will make the difference between victory and defeat. Consider it blasphemy if you wish, but if you use your forces improperly, leadership will simply determine how long you stay in the fight, not the ultimate outcome. So, the student of war who is unversed in the art of war of varying his plans, even though he be acquainted with the Five Advantages, will fail to make the best use of his men. (VIT 6) Though an obstinate fight may be made by a small force, in the end it must be captured by the larger force. (ABS 10)

Chapter 4: DECISION MAKING FOUNDATION
Expansion and Analysis; Part 1

The next step in the reorganization of *The Art of War* into the *Pattern of Power* is the expansion and analysis of our priorities. Sun Tzu's statement, if you know the enemy and know yourself, your victory will not stand in doubt; if you know Heaven and know Earth, you may make your victory complete, (T 31) encapsulates every priority we objectively analyze. I use it as the central principal around which I have organized this section.

In this chapter we will look at the overall process by which we will be able to know ourselves, our enemy and heaven and earth. I have drawn the phases of a campaign from *The Art of War* and put them into the proper order. I have expanded the **Analysis** phase to show you the process we will use to determine the possible COAs that Sun Tzu himself recommends.

Sun Tzu exposed nine different phases of a campaign. They are;
1. Receive Commands. (M 1)
    a. Friendly Mission: What, who, when, where, why (Friendly Mission – Chapter 4; Priorities Part 3)
2. **Analysis** (LP)
    a. Perform Earth Analysis (Terrain and Ground – Chapter 5; Expansion and Analysis Part 2)
    b. Perform Enemy Analysis (Enemy – Chapter 6; Expansion and Analysis Part 3)
    c. Determine strength ratio (Strength – Chapter 7; Expansion and Analysis Part 4)

d. Choose best set of Actions (Possible Actions – Chapter 8; Planning Part 1)
   e. Perform Heaven Analysis (Heaven – Chapter 9; Expansion and Analysis Part 5)
   f. If we can attack, are the conditions right for the use of attack by fire? (The Attack by Fire – Chapter 10; Expansion and Analysis Part 6)
   g. Develop deception plan based on COA (Deception – Chapter 11; Planning Part 2)
   h. Answer other considerations to determine who will win (Ten Questions – Chapter 12)
3. Collect Army (M 2)
4. Concentrate Forces (M 2)
5. Pitch Camp (M 2)
6. Tactical Maneuvering (M 3)
   a. Turning the devious into the direct, and misfortune into gain (M 3)
7. Strike (E 13)
8. Retire (WPS 10)
9. Consequences (TNS)

The **Analysis** phase is where we focus in order to illuminate how *The Art of War* fits within the larger, multi-criteria decision making process (MCDMP).

The Art of War and this *Pattern of Power* will not give you one, nice, clean answer to your situation. It will however, narrow down your possible actions, from which you must develop alternative COAs and then determine the one that will best fulfill your mission.

There are three main areas of analysis you must work through; Terrain and Ground (Earth), Enemy, and Strength. Each of these areas of analysis will provide you with certain suggested actions based on your situation and mission. These areas of analysis should be considered separately before you combine them and allow them to modify each other.

In one set of analysis, you may find that you are able to fight, while in another area the suggestion is to not move. These are not conflicts as much as they are opportunities to effectively think about the strengths and weaknesses of your own situation. You may be sitting on advantageous ground without the strength to succeed. Fighting would then be disastrous. Through the guidance that these principles give you, combined with your imagination and experience, you will create at least three separate and distinct COAs.

Though it will naturally be performed simultaneously, next in this model you perform an analysis of Heaven to establish whether or not the timing of your plan is optimal. Heaven can perform as a veto or a pass to the other possibilities you have determined. At this point you will begin comparing the viability of COAs.

You are embarking on a journey that will force you to open your eyes to the real challenges associated with your goals. The common practice of skipping detailed planning is most often caused by a desire to avoid facing harsh realities. These obstacles exist whether or not we acknowledge them, and they invariably stifle our actions and deflate our desired results. Planning is not easy. It is, however, a process that will allow you to make smart choices, think innovatively about how to achieve your goals, and which will allow you to apply proven, 2,500 year-old principles to your plans.

In graphic form, the process we will build out during our study is as follows. You can see how the following flowchart corresponds to the Analysis phase of a campaign. The broader MCDMP will be explained later in this work, but it is important for you to see early where it fits in our model.

The ART of WAR: Organized for Decision Making

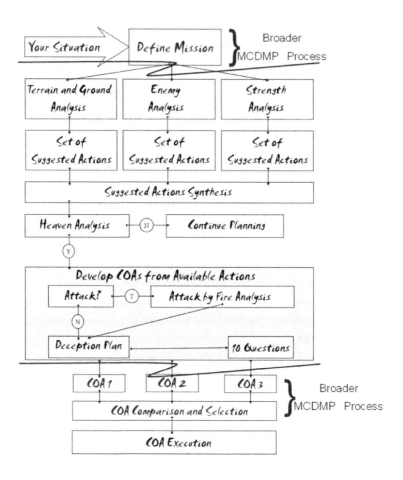

Chapter 5: FRIENDLY MISSION
Priorities; Part 3

In war, the general receives his commands from the sovereign. (M 1/VIT 1) This is the first sentence in the two chapters, Maneuvering and Variations in Tactics. This is the sum total of what is said about mission in *The Art of War*. There is, however, a clear chain of command, from the sovereign to the general to the officers to the soldiers. The sovereign will provide some direction, some desire. Even if it is as broad as, "I want more land so I can become more powerful," or, "I want to increase market share by 27% next year." There is our mission. At least it is a start. The general gets to work determining *how* to make his sovereign more powerful.

The talented general takes Sun Tzu's teachings and starts to figure the best way to accomplish his mission. How do I succeed without fighting? How do I balk the enemy's plans? Which potential enemy can I most easily defeat and how will these actions effect our other neighbors?

You are reading this book with a hope of being able to accomplish something. That is the start of your mission. In order to activate the rest of the *Pattern of Power* you must define that mission. You will refer back to your mission to guide your analysis through remainder of the process. A poor definition results in poor analysis and COA.

As the purpose of this book is to make *The Art of War* practical, powerful and usable for you, I am highlighting this aspect as the starting point from which all of your planning will follow. It is impossible to over emphasize the importance of understanding your mission and objectives. Without a solid foundation in purpose, the reminder of the analysis is useless and potentially dangerous.

A standard mission is defined through the answers to five questions; the 5 W's - what, who, when, where, why. Start with *what* your objective is and it will naturally lead you through the ability to define three of the other four W's. The one that remains unanswered is *why*. You must determine the why behind your objective and examine it for worthiness when held to the light of moral law.

What do you want to accomplish? Though the *what* is not expanded upon in *The Art of War*, your ability to accomplish your mission is covered in the key areas of analysis upon which Sun Tzu does elaborate. With your mission as your guide, you can progress into the other chapters to start clarifying possible COAs, or the *How* of accomplishing your mission.

The analysis of your mission with respect to knowing yourself, your enemy and heaven and earth are roughly coupled in the following way;

1. *Who*, or with what, do you have to accomplish your mission? Study Chapter 8, Strength.
2. By *when* do you need to accomplish your mission? Study Chapter 10, Heaven.
3. *Where* will you accomplish your mission? Study Chapter 6, Terrain and Ground.
4. *Why* do you need to accomplish your mission? Study the Moral Law.

There is a great deal of confusion in the literature regarding the definition of Mission. Though I risk adding to that in the wider sense, for our purposes, your mission is something you want to accomplish. It can be a task that has been handed down from your boss, or it can be a personal or family goal, a group initiative, or team mission. And yes, it can be a company mission statement as well.

Chapter 6: TERRAIN and GROUND (Earth)
Expansion and Analysis; Part 2

The Art of War recognizes nine varieties of ground (TNS 1) and we may distinguish six kinds of terrain (T 1) How to make the best of both strong and weak – that is a question involving the proper use of ground. (TNS 33)

This is the first of three primary areas of analysis you will do to determine an appropriate set of actions. The analysis of Terrain and Ground will provide you with the possible options based on your situation.

We use this step to analyze the *where* part of our mission. Where are you? Where do you want to go? The enemy situation is automatically incorporated into this analysis so a separate look at the enemy with respect to terrain and ground is unnecessary. Depending on which type of ground you find yourself, you should note the possible actions as directed by the situation. These will later be modified by other aspects of your enemy, strength and heaven considerations.

Move to a point of ADVANTAGE

In *The Art of War*, Terrain and Ground are treated as separate conditions. In the *Pattern of Power*, their interconnectivity is recognized and their descriptions are combined. A dissection of the Terrain and Ground reveals four primary categories; Cohesion, Advantage, Disadvantage and

Equality. The fifteen types of terrain and ground are organized as follows;
1. **Cohesion** refers to the condition of your team and asks the question, "Are your people ready for battle?"
2. **Advantage** represents the desired state of your forces with reference to your competitor.
3. **Disadvantage** describes conditions that demand maneuver to a point of advantage.
4. **Equality** indicates those types of terrain and ground that allow no obvious path to advantage, yet do not pose a disadvantage either.

Except for cohesion, the categories will include all aspects of your business and market. Terrain and ground may refer to products, people, money and other resources where you may have or not have an advantage. An analysis of your specific purposes and situations will give you the answers to these questions.

Cohesion

The first three types of ground do not refer to a piece of property or the physical location of your team. These types of ground, Dispersive, Facile and Serious, refer to the development stages of a team and their ability to take on a challenge. Sun Tzu uses phrases like, "inspire my men," "close connection between parts of my army," and "cohesion." These clearly describe the relationship between people, not products, weapons or supplies.

Though your analysis of all of the other terrain and ground information may indicate that you can fight, the condition of your team may tell you otherwise.

1. **Dispersive ground** (TNS 1)

When a chieftain is fighting in his own territory (TNS 2) fight not (TNS 11). I would inspire my men with unity of purpose (TNS 46)

What it means to you. If you take a new or non-cohesive team into battle, there is a high probability that they will fail and disband.

Decision making question. How cohesive is your team? Is your team new? Do you have a new mission? Are there significant discipline problems in your team?

Key to the interpretation. The statement that, "I would inspire my men with unity of purpose," refers to people, not territory. Yes, territory is mentioned, but the important aspect is what that locale means to the unity of the chieftain's army.

Interpretation. Dispersive ground refers to the unity and cohesion of the team. It seems to pertain to a new team or when you have short preparation time. The team is not fully trained, not cohesive and not ready to function at a high level. If the general takes the wrong action and gets his team into a fight, his army will disperse.

Poor discipline, regardless of how long the team has been together, may also be a signal that the team is still on dispersive ground.

The key here is that they are not ready to handle a significant challenge. If you hit them with the big stuff right off the bat, you may blow them out. At the beginning, people are extrapolating what the rest of the mission will look like. If they fight right away, they may choose to bail out early. Take time to get them up to speed until everyone is more committed. Build the team's cohesion and trust. Inspire them by the goal and the moral law, and get them moving first. Develop communication norms.

Keep in mind that this may be ground at the end of a campaign also. The team may be very capable and very cohesive, yet the anticipation of the finish line may make them vulnerable to inattentiveness. The team has achieved some level of success, and they are letting their guard down as they approach their home territory. Maintain the discipline necessary to fully complete the mission. The risk of failure on dispersive ground is that your team may never reach success because the enemy is close and watching for weakened diligence.

2. **Facile ground** (TNS 1)
When an army has penetrated into hostile territory, but to no great distance (TNS 3). Penetrating hostile territory but a short way means dispersion (TNS 42). halt not. I would see that there is close connection between all parts of my army.
What it means to you. Though you have started the mission, be careful of doing too much too fast. If your developing team gets into a big fight there is a high probability that they will fail and disband.
Decision making question. How cohesive is your team? Has your team experienced some early success? Are there signs of role definition and struggles within the team that need to be resolved? Is there a desire to slow down a bit?
Key to the interpretation. The reference to "dispersion" again, and the statement that, "I would see that there is close connection between all parts of my army," refers to people, not territory. Yes, territory is mentioned, but the important aspect is what that locale means to the unity of his army.
Interpretation. Facile ground refers to when a team is still developing. Note that Sun Tzu again uses the word dispersion to describe facile. People are committed as a team but still building their cohesion and capability. Conflict arises within the team as communication norms are developed and responsibilities and authority are tested. There remains the possibility of dispersion if pushed too hard.

You are now in hostile country. Keep going and do the best you can if battle comes your way, but avoid it if possible. By not halting, despite turbulence, your team is forced to deal with and count on each other. Your progress as a leader shows dedication to the mission, which reduces options for the team. The team now has to forage and rely on each other. Constant communication with your team is critical.

A new mission is characterized by an initial burst of energy. Watch for the temptation to back off a little. This is the time to consolidate, refocus and remind everyone of the mission. Place a large goal in front of them that requires their cooperation. Begin looking for increasing efficiencies.

3. **Serious ground** (TNS 1)
When an army has penetrated into the heart of a hostile country, leaving a number of fortified cities in its rear. (TNS 7)
Penetrating hostile territory deeply brings cohesion (TNS 42). Gather in plunder.
What it means to you. Your team is ready to fight.
Decision making question. How cohesive is your team? Are you already deeply committed and with reduced options? Is your team well trained? Do the people on your team trust each other?
Key to the interpretation. The statement describes the territory in terms of cohesion. Having covered much ground as a team and being deep in enemy territory, means that they must be ready for a fight.
Interpretation. This is a more mature team that has overcome some challenges, bonded together, and has committed to the mission. They are ready to quickly deal with emergencies and the natural difficulties that come with a tough mission. They may not be constantly focused on the big picture, but they have developed trust for each other and their leaders.

A companion characteristic of deep penetration into the enemy's territory is the lack of other options available to the team members. As individuals they have no hope of survival, but as an army they can survive. This is the ground upon which you want to fight because of the helpful nature of the conditions to cohesion and order.

By this stage, the army had better be constantly vigilant and as efficient as possible. Squabbles, politics and friction cannot be permitted and should mostly be overcome by now.

Advantage

Advantage represents the desired state of your forces with reference to your competitor. This is the type of ground to which we are moving. In order to achieve victory of any sort, you must occupy one of these types of ground. We seek the advantage in order to achieve a supremely excellent victory. Supreme excellence consists in breaking the enemy's resistance without fighting. [ABS 2)

I reiterate that the loftiest purpose of advantageous ground is not to attack and kill. Even now, you are becoming an advanced strategist so remember; Sun Tzu says that the highest form of generalship is to
- balk the enemy's plans;
- the next best is to prevent the junction of the enemy's forces;
- the next in order is to attack the enemy's army in the field;
- and the worst policy of all is to besiege walled cities. (ABS 3)

4. Ground of intersecting highways (TNS 1)
Ground which forms the key to three contiguous states, so that he who occupies it first has most of the Empire at his command. (TNS 6) When there are means of communication on all four sides (TNS 43) join hands with your allies. (TNS 12) I would consolidate my alliances. (TNS 48) *However*, we cannot enter into alliance with neighboring princes until we are acquainted with their designs. Hence he does not strive to ally himself with

all and sundry, nor does he foster the power of other states (TNS 52)

What it means to you. This is the type of ground that most closely relates to supreme excellence and the highest form of generalship and should therefore be our primary goal to identify and reach first.

Decision making question. What ground in your situation is the ground of intersecting highways?

Key to the interpretation. This is related to supreme excellence because of the statement, "most of the Empire at his command." It is impossible to imagine an empire small enough that one hill or mountain can control the majority of it. This refers to a condition as much as piece of terrain. Since it does not necessarily refer to a location, the meaning of control must be broader than an action like attack or defend. We can control the Empire simply by achieving this condition first.

Interpretation. This is the best choice of all of the ground we are told to reach first. When we have most of the empire at our command, it means much more than some tactical advantage as in accessible ground, narrow passes, and precipitous heights. Through these, we may achieve our objective, but we will probably do it through fighting, destruction and loss.

The key to this COA is the allies we choose and how strongly we are acquainted and aligned with their designs. What incentive have we given them to join us and play a part in our objective? It is implied later that we have spies in these camps also.

Who are your potential allies and for what end? How can you propose agreements with competitors, suppliers, vendors, retailers etc. to squeeze out the competition before they have a chance to fight. This COA takes the most planning and most advance movement. Distance from the enemy is good, but by virtue of the fact that our movements will be known by camps (allies) that we do not control, secrecy will be more difficult if not impossible to maintain. Make sure that each movement is masked by the greater deception in preparation for the anticipated information leak.

Important to keep in mind is the warning against fostering the power of other states. What can you design so you are linked but not fostering?

5. **Contentious ground** (TNS 1)
Ground the possession of which imports great advantage to either side. (TNS 4) Attack not. (TNS 11) I would hurry up my rear. (TNS 47)
What it means to you. Contentious ground is a sub-category of Advantageous ground that consists of accessible ground, narrow passes, and precipitous heights. The suggested COA with regard to each of these types of ground are consistent with "attack not."
Decision making question. Does the ground you wish to posses import a great advantage to you?
Key to the interpretation. Contentious ground is simply described as providing a great advantage to the occupier. This is a condition, not an actual piece of terrain, except as categorized below.
Interpretation. This is the position you will seek to occupy first. It is a commanding position with reference to the place you believe enemy will go, and from where you can achieve a decisive victory, though that victory may not come as a result of combat. The suggestion to not attack is consistent with the order of the highest forms of generalship.

Your purpose in reaching this type of terrain first is not to attack and destroy, which is the third of four choices, but rather to command such an advantage that the enemy's plans are balked.

You will attempt to move your enemy through baits and deception into the area you control. You will not attack from this position, you will stay in this position, and trap the enemy, which may ultimately result in an attack, but better yet will have an enemy retreat or attack your unbreachable position.

a. Accessible ground (T 1)
Which army can occupy accessible ground first and carefully guard its line of supplies?
Ground which can be freely traversed by both sides (T 2) raised and sunny spots, then you will be able to fight with advantage. (T 3)
<u>What it means to you</u>. Accessible ground is advantageous, though due to the ability of both sides to freely traverse it, it is also probably a goal of both sides.
<u>Decision making question</u>. Is there accessible ground in your situation? Can you reach accessible ground first?
<u>Key to the interpretation</u>. The ability of both sides to freely traverse the ground indicates ease, and a path of least resistance.
<u>Interpretation</u>. The suggestion is to *fight*, not *attack*. This is the ground most people look at first as the battle field. This would be the obvious solution and perhaps provide the most ready COA. The general who only prepares for this COA, especially when time is short, will be caught off guard by the general who planned earlier, looked beyond accessible ground to the other two contentious grounds and better yet to the ground of intersecting highways. I would then say, as a COA, achieving the accessible ground first is the forth option.

Rapidity over accessible ground can be assumed to be of the utmost importance. Traversing the ground is easy and can happen quickly. Because of this, the baggage trains may be left behind or trail the main body by a great distance. These supplies could be vulnerable to attack. Be aware of the warning, If asked how to cope with a great host of the enemy in orderly array and on the point of marching to the attack, I should say: "Begin by seizing something which your opponent holds dear; then he will be amenable to your will." (TNS 18)

Your supplies are a key target for the enemy. You may get to the ground first, but if you lose your provisions, you lose. The opportunity here is to include the appearance of seeking the accessible ground as part of your deception plan while you are really moving to one of the other types of Advantageous ground and catching your enemy off guard. If you know that your enemy is seeking the accessible ground first, then you can develop a COA to defeat his strategy.

b. Narrow passes (T 1)
If you can occupy narrow passes first, let them be strongly garrisoned and await the advent of the enemy. (T 8) if you are not first, do not go after him if the pass is fully garrisoned, but only if it is weakly garrisoned. (T 9)

What it means to you. The occupation of ground that controls narrow passes provides a significant advantage that multiplies the effectiveness of your strength.

Decision making question. Are there narrow passes in your situation? Can you reach and control them first?

Key to the interpretation. The definition of narrow passes is a constriction of some type that stretches out the enemy and slows their progress. The goal is to secretly control the pass and allow the enemy to put themselves into a vulnerable situation.

Interpretation. Narrow passes are places where the enemy is constricted. In these situations, enemy movement slows and the focus of the enemy narrows to one bottleneck, which if hampered even further, will severely damage if not destroy the enemy. When looking at COAs, reaching a narrow pass first is being able to take advantage of the constriction of your enemy.

Part of your deception plan may include some enticement for the enemy to enter a narrow pass. When you reach the narrow pass first, your strength is multiplied, and it allows you to consider fighting with a 1-1 ratio. It may make a 5-1 ratio enough to surround the enemy and balk his plans.

The key to the COA would then be how to take advantage of this constriction. While leaving an avenue of escape to the enemy, you must occupy contentious terrain on their route in order to deter or destroy the enemy. It is highly probable that in movement to narrow passes you will pass through difficult or hemmed-in ground.

c. **Precipitous heights** (T 1)
If you can occupy precipitous heights first, wait for him to come up. (T 10) If the enemy has occupied them before you, do not follow him, but retreat and try to entice him away. (T 11)
What it means to you. The occupation of precipitous heights provides a dominant advantage that multiplies the effectiveness of your strength.
Decision making question. Are there precipitous heights in your situation? Can you reach and control them first?
Key to the interpretation. Precipitous heights completely dominate the ground around it. This is a piece of terrain. It is local to a specific objective, unlike ground of intersecting highways which control much more than line of sight.
Interpretation. Precipitous heights are places where you can observe and destroy the enemy from a position of strong leverage. In these situations, the enemy may or may not be constricted, and their movement may or may not be slowed. This is a situation that allows us to thwart the enemy's plans without destroying them. We can deny them a piece of terrain. Our dominant position makes it unwise for them to continue towards our objective. Our tactic may be to simply tap them and force them to take an alternate route, effectively leaving the objective to us. Though our position may command our objective, we may not be able to prevent the escape of the enemy. This would qualify as balking the enemy's plans without destruction, which is the highest form of generalship.

This is a position that even if attacked, will allow us to use stratagem to overwhelmingly defeat the enemy.

Part of your deception plan may include some enticement for the enemy to enter ground that is dominated by precipitous heights. When you reach precipitous heights first, your strength is multiplied, and it allows you to consider fighting with a 1-1 ratio. It may make a 5-1 ratio enough to surround the enemy and balk his plans.

The key to the COA would then be how to take advantage of this domination. While leaving an avenue of escape to the enemy, we must occupy contentious terrain on their route in order to deter or destroy the enemy. It is highly probable that in movement to precipitous heights you will pass through difficult or hemmed-in ground.

d. Entangling ground (T 1)

Ground which can be abandoned but is hard to re-occupy (T 4) you may sally forth and defeat him, but if the enemy is prepared for your coming, and you fail to defeat him, then, return being impossible, disaster will ensue. (T 5)

<u>What it means to you</u>. Entangling ground is the least desirable of Advantageous ground. It entails great risk in the set up and even in the best case only provides the opportunity for a lower category of action; the attack.

<u>Decision making question</u>. Do you have limited opportunities for higher generalship, yet still have the possibility of taking offensive action from a secure and secret place?

<u>Key to the interpretation</u>. The use of the word *attack* instead of *fight*.

<u>Interpretation</u>. Entangling ground is the ONLY terrain and ground where an attack is specifically suggested. The other contentious grounds extend to the point of saying that you can *fight* with advantage, otherwise they say, *guard, wait* and *await*. They do not suggest that you use your advantage by attacking.

This is in keeping with the idea that from contentious ground, our location remains secret and that our actions are from a point of extreme advantage so the outcome is never in question.

Entangling ground does not allow an incredible advantage in itself, so we must revert to a less advantageous form of combat, the attack, and it must be on an unprepared enemy. Attacking the enemy's army in the field is the third of four levels of generalship.

This type of ground takes time to occupy. It takes resources and comes at a cost. It is a situation where you must take some action and invest some effort and you must do it in secret otherwise the enemy will be prepared for you and disaster will ensue.

Many technologies, products, and systems take time and effort to perfect and may meet the definition of entangling ground.

Be aware that even if you occupy entangling ground properly and in secret, your attack may still fail if you do not pay attention the other COA analysis areas. If your attack does not succeed, your investment is lost and the opportunity cost is lost.

Disadvantage
These are types of ground that demand maneuver to a point of advantage in order to avoid disastrous outcomes.

6. **Open ground** (TNS 1)
Ground on which each side has liberty of movement (TNS 5) do not try to block the enemy's way (TNS 12)
I would keep a vigilant eye on my defenses (TNS 48). In dry, level country, take up an easily accessible position, so that the danger may be in front, and safety lie behind. (AOM 9) Utilize the natural advantages of the ground. (AOM 13)

<u>What it means to you</u>. This is a condition, which allows you to maneuver to Advantageous ground, though you have to always keep an eye on your enemy.
<u>Decision making question</u>. Do both you and your enemy have many COA options?
<u>Key to the interpretation</u>. *Liberty of movement*
<u>Interpretation</u>. When options are many for both sides and things can happen quickly, maintain diligence in protecting yourself, but continue movement towards your choice of Advantageous ground. Build your strength if you need to, gather intelligence, and stay nimble.

7. **Difficult ground** (TNS 1)
Mountain forests, rugged steeps, marshes and fens – all country that is hard to traverse (TNS 8) keep steadily on the march (TNS 13). When in difficult country, do not encamp. (VIT 2) Pass quickly over mountains (AOM 1)

Do you have to cross a salt-marsh? Your sole concern should be to get over them quickly, without any delay. (AOM 7) Country in which there are precipitous cliffs with torrents running between, deep natural hollows, confined places, tangled thickets, quagmires and crevasses, should be left with all possible speed and not approached. (AOM 15) While we keep away from such places, we should get the enemy to approach them; while we face them, we should let the enemy have them on his rear. (AOM 16)
<u>What it means to you</u>. When you are in the middle of difficult ground, do not stop to figure things out. Hold the banners high for everyone to follow and keep on course.
<u>Decision making question</u>. Are you in difficult ground or can you predict difficult ground on the way to Advantageous ground? Do you have a goal on which everyone can focus?
<u>Key to the interpretation</u>. *Keep steadily on the march.* This teaching is counter to our natural inclination to stop and find an

easier way when times are tough. Sun Tzu expects that you planned the movement through the difficult ground and therefore have taken the best route and have a higher purpose for doing so. Interpretation. Situations that provide no combat advantage in themselves except that they are necessary to pass over or through in order to get to Advantageous ground. You should expect this type of ground in all cases, but especially when you are planning on occupying narrow passes and precipitous heights.

This is the opposite of contentious ground. If we are engaged by an enemy occupying contentious ground, then we are at a great disadvantage. We must know the current situation on the ground before we put our team in such a position of vulnerability.

Maintain focus on the ultimate objective. Maintain strength and momentum. Through the proper traversing of such difficult ground you may provide an advantage for yourself by arriving at Advantageous ground from an unexpected direction. The enemy naturally does not expect you to cross difficult ground and your deception plan should reinforce their belief.

8. Hemmed-in ground (TNS 1)

Ground which is reached through narrow gorges, and from which we can only retire by tortuous paths, so that a small number of the enemy would suffice to crush a large body of our men (TNS 9) resort to stratagem (TNS 14) I would block any way of retreat. (TNS 50) In hemmed-in situations, you must resort to stratagem. (VIT 2) Do not linger in dangerously isolated positions (VIT 2)

What it means to you. Do not go to hemmed-in ground unless you absolutely have to.

Decision making question. Do you have liberty of movement in several directions or are you in hemmed-in ground?

Key to the interpretation. The statement, *ground which is reached* indicates that you have already traversed the difficult ground, such as narrow gorges, to arrive at hemmed-in ground. Any route of exit from the hemmed-in ground will put you in an extremely vulnerable position.

Interpretation. Situations that allow for only very dangerous options. Traversing this ground must be central to reaching Advantageous ground or a part of our deception plan. If we engage in battle on hemmed-in ground, we must be victorious or the results will be disastrous. Both arriving to, and retreating from, hemmed-in ground means traversing difficult ground. If we are engaged by an enemy occupying contentious ground during that time, then we are at a great disadvantage. Block any retreat your team may have so their focus remains only on victory. In the case of battle, the ground becomes desperate ground.

9. **Desperate ground** (TNS 1)

Ground on which we can only be saved from destruction by fighting without delay (TNS 10) fight. (TNS 14) I would proclaim to my soldiers the hopelessness of saving their lives. (TNS 50) In desperate position, you must fight (VIT 2)

What it means to you. If you are in desperate ground it means that your planning and intelligence failed you.

Decision making question. Where are the places in your plan that have the greatest risk of creating a situation of desperate ground? What are your contingencies in those situations?

Key to the interpretation. The best hope is savior from destruction, not victory.

Interpretation. Desperate ground is not so much ground as a situation. It could be any of the types of ground or terrain. You will only be in this situation when you have not prepared well enough, by knowing yourself, your enemy and the ground. The only option is to get the most out of your team by giving them no other options but to fight. Hence the command, When you surround an army, leave an outlet free. Do not press a desperate foe too hard. (M 36)

This is not a COA choice that we would plan on, though risks can be identified and contingencies planned. Fighting does not imply panic as much as it does determination. This is a no-holds-barred situation. Are there tactics of great risk that you would otherwise not employ? Is there a tactic that will damage the enemy in a massive way but which is also guaranteed to inflict significant loss to your forces?

The suggestion to fight instead of retreat implies movement forward, not backward. How can you create the ability to flee by fighting? If you flee, you may be able to save some forces, but you have ultimately lost.

Equality
These are types of ground that allow no obvious path to advantage, yet do not pose a disadvantage either.

10. Temporizing ground (T 1)
When the position is such that neither side will gain by making the first move, it is called temporizing ground. (T 6) Even though the enemy should offer us an attractive bait, it will be advisable not to stir forth, rather to retreat, thus enticing the enemy in his turn; then, when part of his army has come out, we may deliver our attack with advantage. (T 7) Move not unless you see an advantage; use not your troops unless there is something to be gained; fight not unless the position is critical. (ABF 17) He will win who knows when to fight and when not to fight. (ABS 17)

What it means to you. This is a situation where relative strength is equal. The first move should not be to engage the enemy, but rather in a direction towards Advantageous ground.

Decision making question. Are you occupying temporizing ground? How can you get to Advantageous ground and lure the enemy?

Key to the interpretation. The statement that *neither side will gain by making the first move* indicates that for whatever reason, the combination of terrain, ground and resources create an even-strength ratio.

Interpretation. This is a situation where the general must know when to fight and when not to fight. In pursuit of your objective under these circumstances, it is unwise for you to take aggressive action or the first move. You may have arrived in this situation after a great deal of effort and expense and you are emotionally prepared for battle. This is not the time to fight. Recognition of this fact gives you an advantage over your enemy who may not posses the same control.

The suggestion *to retreat as an enticement* plays upon the emotion of your enemy. Quickly get back to planning and improving your situation as you move to Advantageous ground. What do not you know?

f.   **Positions at a great distance from the enemy.** (T 1)
If you are situated at a great distance from the enemy and the strength of the two armies is equal, it is not easy to provoke a battle, and fighting will be to your disadvantage (T 12)

What it means to you. Not enough information on the enemy to determine what ground is Advantageous. Fortunately, the conditions are such that the enemy does not have an advantage over you either.

Decision making question. Is there too much distance between you and your goal to determine how the enemy may react? Does the enemy have a current advantage over you?

Key to the interpretation. Great distances mean that you have little information on which to determine a move.

Interpretation. You have an objective that is far in the future. It will take a long time or a lot of effort, resources and chances to reach, and you are not fully aware of your enemy's intentions. You must figure out through intelligence what your enemy's intentions are before you know how close you are to them.

This is not so much about who can reach advantageous ground first, but finding out what advantageous ground actually is. Will your preemptive moves mean anything? Can you reach the ground first only to find that the enemy has gone the other way? How fast can you close the distance? If you close the distance will you be able to occupy advantageous ground first?

As distances go, they must be taken into consideration with the other types of ground, especially the three that relate to the development of the team. It seems reasonable that you will be closer to the enemy when you are deep in his territory and farther away when in yours. Of course, if they are attacking you, then the opposite would apply.

It is understood in this type of ground that relative strength is equal and therefore the enemy does not have an advantage over you and cannot gain one very quickly.

Chapter 7: ENEMY
Expansion and Analysis; Part 3

This is the second of three primary areas of analysis you will do to determine an appropriate set of actions. The analysis of the Enemy will provide you with the possible options based on your situation. In Terrain and Ground we started this process because many of the questions forced us to take a look at our circumstances relative to the enemy.

The enemy analysis is a critical step in determining what we know and what we still need to learn about our enemy. Not only will you break down the enemy situation, but you will then analyze how to get the missing information through a look at the use of spies.

Depending on the enemy situation, you should note the possible actions as directed by Sun Tzu. Other aspects of your terrain and ground, strength and heaven considerations will later modify these.

Knowing your Enemy
How well do you know the enemy situation? 0% - 100%. How well can you answer the 5 W's of a mission statement on your enemy?
- *What* does your enemy want to accomplish? Be specific. "Clear and concise"
- *Who* does your enemy have to accomplish its objectives with? Strength.
- *Where* will your enemy accomplish it? Ground and Terrain.
- By *when* does your enemy need to accomplish it? Time.

- *Why* does your enemy need to accomplish it? Moral Law.

The 5 W's are a way to organize your information gathering. Depending on what you find, the following are the suggested actions from *The Art of War* with respect to the enemy's actions.

Suggested Actions
If the enemy is taking his ease, he can harass him;
If the enemy is well supplied with food, he can starve him out;
If the enemy is quietly encamped, he can force him to move. (WPS 4)
When the enemy is close at hand and remains quiet, he is relying on the natural strength of his position. (AOM 18) **Be careful of making concessions or alterations in your plan when you have limited information on the enemy and they seem steadfast in their defense or position. They may be stronger than you estimated. Look for deceptions.**
A clever general, therefore, avoids an army when its spirit is keen. (M 29)
A clever general attacks an army when it is sluggish and inclined to return. (M 29)
- When the soldiers stand leaning on their spears, they are faint from want of food (AOM 29)
- If the officers are angry, it means that the men are weary. (AOM 33)
- If those who are sent to draw water begin by drinking themselves, the army is suffering from thirst. (AOM 30)

What signs are there that the enemy is tired. Have they been working long hours for extended periods and not made the progress they need to. Do they still have a long road and lots of work still ahead? Realize that they may still be motivated though they are tired. Are their resources running out? Has turnover begun to rise and funds dwindle? Are they sacrificing normal raises or extra benefits? Are they working on a shoestring budget? Have you worked with a negotiating party from the enemy that has started to see if there is something in it for themselves?

Await the appearance of disorder and hubbub amongst the enemy (M 30)

- Clamor by night betokens nervousness. (AOM 32) **Emergency meetings. Sudden layoffs.**
- If there is disturbance in the camp, the general's authority is weak. (AOM 33)
- If the banners and flags are shifted about, sedition is afoot. (AOM 33) **Change in leadership, demotions, layoffs, shift in org chart.**
- The sight of men whispering together in small knots or speaking in subdued tones points to disaffection amongst the rank and file. (AOM 35) **High turnover.**

Do not pursue an enemy who simulates flight (M 34)

When some are seen advancing and some retreating, it is a lure. (AOM 28) **The draw of a retreating enemy is tempting. When the enemy divides its force, and moves in two different directions, even if one seems to be of no danger to you, a trap is set. Remain in your advantageous position.**

Do not swallow bait offered by the enemy. (M 35)

- Peace proposals unaccompanied by a sworn covenant indicate a plot (AOM 26) **Insincerity. Flattery. False Praise. Uncharacteristically meek demeanor. Inconsistencies. Paradoxical movements.**
- When he keeps aloof and tries to provoke a battle, he is anxious for the other side to advance. (AOM 19) **Arrogance and prompting. Daring or challenging.**
- If his place of encampment is easy of access, he is tendering a bait. (AOM 20) **Probably on Accessible ground. Often it is the only ground analyzed by the general and they fall into the trap of *apparent* advantage.**

If there are more defendable encampments nearby and the enemy chooses an easily accessible one, then an ambush is set. When an organization seemingly capitulates, or fails to put up a good fight for something of seeming value, there is an ambush.

Do not interfere with an army that is returning home. (M 35)
When you surround an army, leave an outlet free. (M 36)
Do not press a desperate foe too hard. (M 36)

- Too frequent rewards signify that the enemy is at the end of his resources (AOM 36)
- Too many punishments betray a condition of dire distress. (AOM 36)
- If the enemy sees an advantage to be gained and makes no effort to secure it, the soldiers are exhausted. (AOM 31) When an organization does not take an expected COA, there is some reason for it. Look for explanations as to why they did not take an obvious advantage. Perhaps they do not have the resources or strength you estimated.
- When envoys are sent with compliments in their mouths, it is a sign that the enemy wishes for a truce. (AOM 38) A sudden turn in the attitude of a competitor towards the positive may indicate a change in strategy. Look for other signs of intent.

Refrain from intercepting an enemy whose banners are in perfect order. (M 32)
Refrain from attacking an army drawn up in calm and confident array. (M 32)
Do not advance uphill against the enemy (M 33)
Do not oppose the enemy when he comes downhill (M 33)

What signs are there that an attack is coming?

- When an army feeds its horses with grain and kills its cattle for food, and when the men do not hang their cooking-pots over the camp-fires, showing that they will not return to their tents, you may know that they are determined to fight to the death. (AOM 34) **What investments are made by your enemy in personnel and money? What signs are there of the level of commitment of your enemy? Is there an overwhelming expenditure, that if unsuccessful, their organization dies? You need spies to determine this since the information is on the inside.**

- When there is much running about and the soldiers fall into rank, it means that the critical moment has come (AOM 27) **This is when an organization gets to the point where they no longer feel compelled to hide their intentions. Assume that all actions are deceitful until the point that they start to lock and load.**
- Humble words and increased preparations are signs that the enemy is about to advance. (AOM 24)

Conversely, violent language and driving forward as if to the attack are signs that he will retreat. (AOM 24)

If the enemy's troops march up angrily and remain facing ours for a long time without either joining battle or taking themselves off again, the situation is one that demands great vigilance and circumspection. AOM

Additional considerations when trying to figure out the enemy's plan
What signs are there that the enemy is advancing? Or not advancing? (AOM)
What signs are there that the enemy wants to make us suspicious? (AOM)
What other signs are there that indicate the enemy's disposition? (AOM)

Within this chapter on Enemy Analysis is included direction on the use of spies because the sole purpose of spies is to better know your enemy. Spies are the best means by which to gain information on your enemies.

The Use of Spies
Spies are a most important element in war, because on them depends an army's ability to move. (UOS 27)
What enables the wise sovereign and the good general to strike and conquer, and achieve things beyond the reach of ordinary men, is foreknowledge. (UOS 4)
This foreknowledge;
- cannot be elicited from spirits
- cannot be obtained inductively from experience
- cannot be obtained by any deductive calculation (UOS 5)
- of the enemy's dispositions can only be obtained from other men (UOS 6)

There are five classes of spies
1. Local spies. (UOS 7) Employing the services of the inhabitants of a district (UOS 9) **Inhabitants of the district are everyone that immediately surrounds the enemy. It is their country, their market, their competitors, the people and organizations that they routinely interact with or pass in their daily activities. These people may or may not have an interest in your enemy's success. Their usefulness is in the information they can provide that may or may not confirm what you get from other sources. They probably do not have specific or direct information of the depth you are looking to get.**
2. Inward spies. (UOS 7) Making use of officials of the enemy (UOS 10) **Officers and soldiers who work inside the organization. These are people who know the secrets and inside information you want to know. These are employees. They are working on the projects that will affect you and your army at some point.**

3. Converted spies (UOS 7)
   - Getting hold of the enemy's spies and using them for our own purposes. (UOS 11)
   - The enemy's spies who have come to spy on us must be sought out, tempted with bribes, led away and comfortably housed. (UOS 21)
   - It is through the information brought by the converted spy that we are able to acquire and employ local and inward spies. (UOS 22)
   - It is owing to his information, again, that we can cause the doomed spy to carry false tidings to the enemy. (UOS 23)
   - Lastly, it is by his information that the surviving spy can be used on appointed occasions. (UOS 24)

   In today's environment, where legal issues deter organizations from incorporating spies, there are fewer of these. Add to that the difficulty of finding them, and I am not sure this has much of a place. However, what about people who used to be inside of the organization? The key is that we are looking for current information that will allow us to use the other types of spies more effectively.

4. Doomed spies. (UOS 7) Doing certain things openly for purposes of deception, and allowing our spies to know of them and report them to the enemy (UOS 12) **This is a key part of the deception plan and should be included with every plan. Are there people we can use to pass false information to the enemy? What are other sources for passing false information or giving a deceptive appearance. Newspapers, reports, etc.**

5. Surviving spies. (UOS 7) Are those who bring back news from the enemy's camp (UOS 13) **These are the people that actually contact our camp to let us know what has been discovered. These are the people who manage the contact list of other spies. These are the people who bridge the gap, over**

dangerous territory and potentially hazardous check points to transport the information. Chance of discovery is always there and the price for discovery is high. You must have these spies otherwise the information will sit where it is and do you no good in the planning process.

Legally gaining copious amounts of foreknowledge should be a high priority for any organization in a competitive environment. What enables the wise sovereign and the good general to strike and conquer, and achieve things beyond the reach of ordinary men, is foreknowledge. (UOS 4) The converse of this statement is obvious and unfortunately, typically ignored.

> **Though the enemy be stronger in numbers, we may prevent him from fighting. Scheme so as to discover his plans and the likelihood of their success. (WPS 22)**

Dedicate an entire organization to this task. Dedicate never before seen amounts of resources to this mission. Be innovative and flexible and you will gain insights from legally obtained information like you cannot believe.

Use the Army as an example again. The Army does not have spies in the enemy brigade on the other side of the hill. What they do have is a dedicated intelligence department that knows and tracks the enemy. They have templates based on their knowledge, which can be updated quickly, and which identifies critical pieces of information that are needed in order to confirm or deny certain enemy courses of action.

Chapter 8: STRENGTH
Expansion and Analysis; Part 4

This is the third of three primary areas of analysis you will do to determine an appropriate set of actions. The analysis of the Strength will provide you with possible options based on your situation. Through the definition of your mission and the analysis of the enemy, you have already compiled the information necessary to determine the Strength ratio. Strength analysis is a subjective yet critical exercise. Carefully compare the opposing army with your own, so that you may know where strength is superabundant and where it is deficient. (WPS 24) The amount of subjectivity varies based on the amount of accurate information you have on the enemy's resources and dispositions. In the best case you have perfect information on the enemy's resources; hard numbers, locations and types. Even then, you will apply judgments to the motivation, capabilities and determination of their forces as they relate to yours.

In *The Art of War*, very clear guidance is provided based on the ratios you determine. These ratios set parameters for the types of actions to take and must be taken into account with the other areas of analysis, Enemy, Terrain and Ground and Heaven.

The standards are cut and dry. When you look closely, however, you will notice some caveats that focus the analysis of strength at a critical point on the battlefield. The determination of which point that is, and your analysis necessitate the gathering of information, detailed planning and careful consideration of all pertinent facts.

Strength ratio

Ten to the enemy's one, to surround him (ABS 8)
- In the practical art of war, the best thing of all is to take the enemy's country whole and intact; to shatter and destroy it is not so good. (ABS 1)

Five to one, to attack him (ABS 8)
Twice as numerous, to divide our army into two (ABS 8)
Equally matched, we can offer battle (ABS 9)
- What we can do is simply to concentrate all our available strength. (AOM 40)
- Keep a close watch on the enemy, and obtain reinforcements. (AOM 40)

Slightly inferior in numbers, we can avoid the enemy (ABS 9)
Quite unequal in every way, we can flee from him (ABS 9)

Additional Insights

Rouse him, and learn the principle of his activity or inactivity. Force him to reveal himself, so as to find out his vulnerable spots. (WPS 23) We can form a single united body, while the enemy must split up into fractions. Hence there will be a whole pitted against separate parts of a whole, which means that we shall be many to the enemy's few. (WPS 14) Numerical weakness comes from having to prepare against possible attacks; numerical strength, from compelling our adversary to make these preparations against us. (WPS 18)

Though an obstinate fight may be made by a small force, in the end it must be captured by the larger force. (ABS 10)

Strength Comparison

What would each of your strengths be on temporizing ground? If there were no advantages provided by heaven and earth considerations, what would your strengths be?

Determine your strength ratio by dividing your strength value by the enemy's strength value.

Friendly	Quantity	Enemy	Quantity
# people		# people	
# people in each critical specialty		# people in each critical specialty	
Relative power of specialty (1x = 5y)		Relative power of specialty (1x = 5y)	
# systems		# systems	
# critical systems		# critical systems	
Relative power of systems (1x = 5y)		Relative power of systems (1x = 5y)	
Amount of money to spend		Amount of money to spend	
Product strength		Product strength	
Allies and their strength		Allies and their strength	
Potential allies and their strength		Potential allies and their strength	
Target allies and their strength		Target allies and their strength	
Subjective judgment on strength for both you and your enemy. 1 is weakest, 10 is strongest.			

Friendly strength value / Enemy strength value

Ratios
- 10 means 10 to 1 advantage
- 5 means 5 to 1 advantage
- 2 means 2 to 1 advantage
- 1 means 1 to 1 equally matched
- .8 means you are slightly inferior in numbers
- .5 means you are quite unequal in every way

No guidance is given to the exact effect that Ground and Terrain make to the strength ratio. I suggest that you do not make decisions based on what-ifs or based on terrain you do not yet occupy. However, it is important to do a strength ratio analysis for the advantageous ground you seek to occupy in order to know what opportunities may or may not present.

It makes sense that advantageous ground can increase the effect of your available strength. For general guidance, the U.S. Army considers it necessary to have a 3:1 advantage in order to attack. If you can get to advantageous terrain first, how will that affect your strength ratio? What are the risks do your forces incur by moving to advantageous terrain?

Chapter 9: POSSIBLE ACTIONS
Planning; Part 1

Let's now earnestly start our planning and move into decision making. In this chapter we are going to summarize the possible actions that Sun Tzu proposes within each area of analysis that we have already conducted. We will explore how these suggestions connect to and influence each other.

Some people may want to argue that this detailed method of planning is useless for war and business because those are such fluid environments. They wonder how they can choose a COA and expect that it will remain the same throughout the operation. They believe it is naïve to spend so much time planning and researching. These are the people who have never seen the value of detailed planning and are in the habit of following their natural inclination of FIRE-READY-AIM.

Those prone to this habit will misinterpret Sun Tzu's statement, "for it is precisely when a force has fallen into harm's way that is capable of striking a blow for victory," (TNS 59) to mean that you have got to get into the mix before you can really use *The Art of War* to accomplish victory, or that your detailed plans will be useless once you get onto the battlefield and see what the real situation is.

It is here that I remind you that Sun Tzu also says, In all fighting, the direct method may be used for joining battle, but indirect methods will be needed in order to secure victory. (E 5) Flexibility on the battlefield is acknowledged, but it must be planned for. The general who wins a battle makes many calculations in his temple ere the battle is fought. (LP 26) This is the general that shapes the battlefield beforehand to meet his desires, and who is familiar enough with the enemy to predict his moves and prepare for a series of indirect maneuvers.

If you want to jump right in without detailed, concentrated planning you cannot answer in the positive, *any* of the considerations that predict victory or defeat. You will fail. You cannot fully utilize these principles for tomorrow's meeting or next week's battle.

Before we go too far, let's examine again what Sun Tzu says is the highest pursuit of the general.

Thus the highest form of generalship is to
1. balk the enemy's plans;
2. the next best is to prevent the junction of the enemy's forces;
3. the next in order is to attack the enemy's army in the field;
and the worst policy of all is to besiege walled cities. (ABS 3)

Now, your job is to apply wisdom and the guidelines compiled from our analysis, then based on your specific situation to form a plan. Your job as a leader is to take these prescribed actions and combine them in a way that produces a COA that accomplishes your mission.

You should see now how the suggested actions in each area of analysis play on each other. Strength says divide our army into two, Terrain and Ground says await, and the Enemy says refrain from attacking. What does it mean and how do you apply the process?

This is not easy, and your plan will contain some assumptions and predictions. These will be less subjective the better your spies ...uh hmmm...your information-gathering systems work. As you sit at your desk, you must foresee the phases as your army moves towards advantageous ground and victory.

Ignorance may be bliss, but knowledge provides a path to success. The multi-criteria decision making process provides a framework for organizing your knowledge and determining what you still need to learn. The complexity of the task at hand, even after the reorganization of *The Art of War*, gives you an idea of how difficult it was to practically apply Sun Tzu's philosophy prior to this reorganization for decision making. It was impossible to determine priorities and suggested actions. Prior to the *Pattern of Power*, the proper and effective application of *The Art of War* was a guessing game.

Strength Ratio	COA
ten to the enemy's one	surround him
five to one	attack him
twice as numerous	divide our army into two
equally matched	we can offer battle
slightly inferior in numbers	avoid the enemy
quite unequal in every way	flee from him

Terrain/Ground	COA
dispersive ground	fight not
facile ground	halt not
serious ground	gather in plunder
contentious ground	attack not
accessible	fight
narrow passes	await
precipitous heights	wait
open ground	defend
ground of intersecting highways	join hands with allies
difficult ground	steadily march
hemmed-in ground	resort to stratagem
desperate ground	fight
entangling	attack
temporizing	retreat then attack
great distances	do not fight

In general, if strength suggests that you can fight, and you are not on ground where combat is recommended, then the continuation of the COA is to move to Advantageous ground, where battle is suggested. As COAs begin to crystallize for you, remember to refer to the guidance of your mission and purpose.

COA Enemy modifiers	COA
taking his ease	attack not harass
well supplied with food	starve out
quietly encamped	force to move
spirit keen	do not attack
sluggish	attack
disorder and hubbub	await
simulating flight	do not pursue
offering bait	do not swallow
returning home	do not interfere
desperate	do not press too hard
banners in perfect order	refrain from intercepting
drawn up in calm and confident array	refrain from attacking
determined to fight to the death	prepare for attack
remain facing ours	great vigilance and circumspection
is up hill	do not advance against
coming downhill	do not oppose

One obstacle that immediately presents itself is the sequencing of the COAs. You may start in facile territory, but quickly move to hemmed in and then accessible ground. How do you plan for that? Incorporate a phased plan.

> Avail yourself also of any helpful circumstances over and beyond the ordinary rules. And according as circumstances are favorable, one should modify one's plans. (LP 16)

Full-Spectrum Decision Making
This is the proper time to more fully introduce Full-Spectrum Decision Making (FSDM). FSDM is a customized multi-criteria decision making process (MCDMP). FSDM is a structured approach to creating innovative solutions to complex, information-rich and multi-objective problems. More importantly, it is a system that can be fully exploited with a pencil and paper by leaders at all levels. Like most topics, the MCDMP has been complicated to the point of uselessness by academics and consultants trying to be smarter than each other or to create a process that requires their *expert* advice. The result is a set of computer programs that take an advanced degree to understand. We reject the complicated for the beauty of effective simplicity.

The FSDM method allows people to;
- See what others cannot see
- Handle massive amounts of information
- Make the complex understandable
- Be decisive
- Effectively engage their imagination
- Lead and communicate with a team

FSDM includes problem analysis, Development of COAs, COA comparison, sensitivity analysis and COA recommendation and decision.
People who attempt decisions for complex issues without a method are at a severe disadvantage in terms of probabilities for success. The vast majority of managers, 99.9% I would say, when faced with a problem inappropriately jump

right into the COA development stage. They overwhelmingly rely on a small set of facts, unaware of the effects of their multitudes of assumptions on their choice solution, and rely on experience and intuition to guide their actions. The performance of their choice is undoubtedly highly sensitive to unknown and unsubstantiated assumptions, making them fragile and vulnerable.

Sun Tzu admonishes us to plan extensively. It can be concluded from his writings that our general will have considered in depth several different COAs. Our normal manager typically only marginally considers all but one COA. The fact that even after all of our analysis, Sun Tzu does not give us one hard and fast COA, and that multiple siloed considerations interact with each other in the end, is consistent with today's best complex decision making methods.

Deep analysis is needed first to define the problem and determine a small set of guiding objectives. Then we enter the COA development step of the FSDM process to use Sun Tzu's wisdom to determine parameters for the development of at least three distinct COAs.

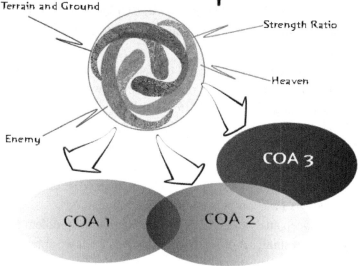

We perform the remainder of the FSDM process to select the most robust and powerful COA. The beauty of the process is revealed in the strategy; a strategy that is derived through concentrated thought and insights that are invisible to the normal person.

Sun Tzu says, All men can see the tactics whereby I conquer, but what none can see is the strategy out of which victory is evolved. (WPS 27) As I have pleaded with you to see this philosophy in total and not take pieces out of context, also know that Sun Tzu says to plan in detail. The general who wins a battle makes many calculations in his temple ere the battle is fought. (LP 26) This means that the serious student of success will take Sun Tzu as a whole and see that its analysis fits as a piece in the greater whole of FSDM complex problem solving.

Notice that I have included Heaven in the above diagram but have not yet presented it as a topic of analysis. Sun Tzu expands very little on Heaven and makes no direct suggestions with regards to specific conditions. Sun Tzu asks as part of the ten questions, with whom lie the advantages of heaven and earth. (LP 13) Though we know the importance of heaven and we know that it must be incorporated into our COA development, the suggestions that derive from its analysis are broad and naturally underlie some of the other areas of analysis. Heaven is covered in the next chapter.

Additional Considerations

This is the future past of *The Art of War* and ties to planning and strategy.

Sun Tzu said: The good fighters of old first put themselves beyond the possibility of defeat, and then waited for an opportunity of defeating the enemy. (TD 1) To secure ourselves against defeat lies in our own hands, but the opportunity of defeating the enemy is provided by the enemy himself. (TD 2) Thus the good fighter is able to secure himself against defeat, but cannot make certain of defeating the enemy. (TD 3) He wins his battles by making no mistakes. Making no mistakes is what establishes the certainty of victory, for it means conquering an enemy that is already defeated. (TD 13) Hence the skillful fighter puts himself into a position which makes defeat impossible, and does not miss the moment for defeating the enemy. (TD 14) How victory may be produced for them out of the enemy's own tactics – that is what the multitude cannot comprehend. (WPS 26) Thus it is that in war the victorious strategist only seeks battle after the victory has been won, whereas he who is destined to defeat first fights and afterwards looks for victory. (TD 15) In all fighting, the direct method may be used for joining battle, but indirect methods will be needed in order to secure victory. (E 5) After that, comes tactical maneuvering, than which there is nothing more difficult. The difficulty of tactical maneuvering consists in turning the devious into the direct, and misfortune into gain. (M 3) For it is precisely when a force has fallen into harms way that it is capable of striking a blow for victory. (TNS 59)

So in war, the way is to avoid what is strong and to strike at what is weak. (WPS 30)

How to make the best of both strong and weak – that is a question involving the proper use of ground. (TNS 33) With his forces intact he will dispute the mastery of the Empire, and thus, without losing a man, his triumph will be complete. This is the method of attacking by stratagem. (ABS 7)

When you engage in actual fighting, if victory is long in coming, then men's weapons will grow dull and their ardor will be damped. If you lay siege to a town, you will exhaust your strength. (WW 2) Again, if the campaign is protracted, the resources of the State will not be equal to the strain. (WW 3) Now, when your weapons are dulled, your ardor damped, your strength exhausted and your treasure spent, other chieftains will spring up to take advantage of your extremity. Then no man, however wise, will be able to avert the consequences that must ensue. (WW 4) Thus, though we have heard of stupid haste in war, cleverness has never been seen associated with long delays. (WW 5) There is no instance of a country having benefited from prolonged warfare. (WW 6)

What the ancients called a clever fighter is one who not only wins, but excels in winning with ease.

(ties to WW #5) (TD 11)

If we wish to fight, the enemy can be forced to an engagement even though he be sheltered behind a high rampart and a deep ditch. All we need do is attack some other place that he will be obliged to relieve. (WPS 11) If we do not wish to fight, we can prevent the enemy from engaging us even though the lines of our encampment be merely traced out on the ground. All we need do is to throw something odd and unaccountable in his way. (WPS 12)

Ponder these compiled statements from *The Art of War* before continuing.
- If you know the enemy and know yourself, your victory will not stand in doubt; if you know Heaven and know Earth, you may make your victory complete. (T 31)
- If you know the enemy and know yourself, you need not fear the result of a hundred battles. If you know yourself but not the enemy, for every victory gained you will also suffer a defeat. (ABS 18)
- If you know neither the enemy nor yourself, you will succumb in every battle. (ABS 18)
- He who exercises no forethought but makes light of his opponents is sure to be captured by them. (AOM 41)
- Success in warfare is gained by carefully accommodating ourselves to the enemy's purpose (TNS 60)
- What enables the wise sovereign and the good general to strike and conquer, and achieve things beyond the reach of ordinary men, is foreknowledge (UOS 5)
- All men can see the tactics whereby I conquer, but what none can see is the strategy out of which victory is evolved. (WPS 27)
- Walk in the path defined by rule, and accommodate yourself to the enemy until you can fight a decisive battle. (TNS 67)
- Move not unless you see an advantage; use not your troops unless there is something to be gained; fight not unless the position is critical. (ABF 17)
- So in war, the way is to avoid what is strong and to strike at what is weak. (WPS 30)

- We may take it then that an army without its baggage-train is lost; without provisions it is lost; without bases of supply it is lost. (M 11)
- To secure ourselves against defeat lies in our own hands, but the opportunity of defeating the enemy is provided by the enemy himself. (TD 2)
- The spot where we intend to fight must not be made known (WPS 16)
- Rapidity is the essence of war (TNS 19)
- All warfare is based on deception. (LP 18)
- Hostile armies may face each other for years, striving for the victory which is decided in a single day. (UOS 2)
- A victorious army opposed to a routed one, is as a pound's weight placed in the scale against a single grain. (TD 19)
- To muster his host and bring it into danger: - this may be termed the business of the general (TNS 40)
- Do not repeat the tactics which have gained you one victory, but let your methods be regulated by the infinite variety of circumstances. (WPS 28)

These are some of the most important principles in the entire book and they should have already prompted your imagination. Here is where we start to organize the philosophy, bring in the expanded teaching and make it relevant to you – **today**.

Chapter 10: HEAVEN
Expansion and Analysis; Part 5

Heaven can be looked at as a series of uncontrollable trends that you can operate in accordance with or try to fight. They will act as a veto of an action and can further modify the COA based on an in-depth analysis. These typically deal with the timing of the COA.

By when do you need to accomplish your mission? Heaven provides us with a logical check on enthusiasm.

Heaven is the environment. It comprises the surrounding trends and conditions such as political, legal, public opinion, market trends, and actual weather (hurricanes, fires, disasters, earthquakes). It can be population trends, area population growth and shrinkage, and other demographics. Whether or not they are advantageous depends on what you want, your capabilities and preparedness for the conditions. If the conditions are night, do you have flashlights, if day, can you hide?

Avoid the extreme trends that are completely against you no matter how good your idea or how compelling your mission. Sun Tzu asks as part of the ten questions, with whom lie the advantages of heaven and earth. (LP 13)

Sun Tzu identifies six characteristics of Heaven;
1. Night (LP 7) **Night obscures movements, but also may make it difficult for us to know where we are going. It is typically colder than day. Is there a full moon, half moon, no moon? What is the cloud cover? When does the moon rise and set?**

2. Day (LP 7) Day movements are more likely to be observed by the enemy. Control of our forces is easier. It is typically warmer than night. When does the sun rise and set?
3. Cold (LP 7) The colder the worse. Consider this a trend that is in the opposite direction of what will help you. Progress will be slow, difficult and costly, and even after long distances; there will be no prospects for speeding up.
4. Heat (LP 7) The hotter the worse. Consider this a trend that is in the direction that will help you. The hotter it is the more mature the trend and it may be on a bubble, or at the top of the cycle. If you jump in at the hottest point, you will lose your investment. If you go when it is too hot, it indicates that you have not planned far enough ahead to catch the trend at the beginning, which would make success much easier. You may start off quickly, but you will bog down, or just as bad, grow too fast.
5. Times (LP 7) Where are the trends in their life cycle? What trend is approaching? Day, night, heat, cold.
6. Seasons (LP 7) Seasons are described as a combination of the first four characteristics. For instance, summer is long, hot days and short nights. Winter is short, cold days and long cold nights.

Heaven

Conditions not too extreme	Progress as indicated
Conditions too extreme	Postpone until better conditions

Chapter 11: ATTACK BY FIRE
Expansion and Analysis; Part 6

Attack by fire is a sub-category of attack, which we have established many times over is the third of four levels of generalship. Destroying is a weak choice, however, if you are destined to destroy, then understand that there is a right time, place and way to carry it out.

There are five ways of attacking with fire. (ABF 1)
1. burn soldiers in their camp
2. burn stores
3. burn baggage trains
    i. We may take it then that an army without its baggage-train is lost; without provisions it is lost; without bases of supply it is lost. (M 11)
    ii. Begin by seizing something which your opponent holds dear; then he will be amenable to your will. (TNS 18)
4. burn arsenals and magazines
5. hurl dropping fire amongst the enemy (ABF 1)

Fire is attention getting - bright. Not easily handled by anyone once started. Fire spreads naturally and destroys. Crises are often called fires. How big is the fire? Small camp fires among groups, or major newsworthy scandals, product malfunctions or deception?

The material for raising fire should always be kept in readiness. (APF 2)

The first material is information provided by spies. This information is the core of the damaging issue. Is it a real product issue or scandal? Can the information be used inside of the company to create distrust or discontent? You must have media contacts at the ready. You must have PR outlets, shareholder information outlets, investment information outlets, market contacts, suppliers etc. You must have agents prepared to operate independent of your organization.

The materials can be rumors, lawsuits, TV, 60 Minutes, newspapers, government, IRS, or others.

In attacking with fire, one should be prepared to meet five possible developments: (ABF 5)
1. When fire breaks out inside the enemy's camp, respond at once with an attack from without. (ABF 6) **Create distention within the enemy's camp - distrust, restlessness, and insecurity. The attack may be "coming to the rescue" of the employees or company. Maintain the enemy's company in tact if possible.**
2. If there is an outbreak of fire, but the enemy's soldiers remain quiet bide your time and do not attack. (ABF 7) **If discipline of the organization is strong and trust is high, they may easily control the fire. Any piling on or rescuing could be seen as invading privacy or worse yet, exposing your strategy and the possible involvement in creating the fire.**
3. When the force of the flames has reached its height, follow it up with an attack, if that is practicable; if not, stay where you are. (ABF 8) **The fire destroys, but may be too big to get near. It may create enough damage that there is nothing worth saving. Again, you must be extremely careful because you cannot plan in advance how much you will destroy.**
4. If it is possible to make an assault with fire from without, do not wait for it to break out within, but deliver your attack at a favorable moment. (ABF 9) **The fire is not set within the**

organization, but in a related area that you are hopeful will spread to the nearby enemy.
5. When you start a fire, be to windward of it. Do not attack from the leeward. (ABF 10) **Your calculations must be extremely conservative and safe because if you are in a wrong relative position, the fire will damage or destroy you too. And once it is set, the wind moves the fire and you do not control the wind.**

There is a proper season for making attacks with fire, and special days for starting a conflagration. (ABF 3) **Much of this depends on Heaven. Depending on how you define and analyze heaven, what are the criteria that make fires appropriate?**

The proper season is when the weather is very dry; the special days are those when the moon is in the constellations of the Sieve, the Wall, the Wing or the Cross-bar; for these four are all days of rising wind. (ABF 4) **Dry weather and rising wind make starting fires appropriate. This takes into account trends of nature. For a fire to be started by an outside source, some conditions that make it more effective would be similar issues or problems in the news, or known problems within the enemy's ranks.**

In the practical art of war, the best thing of all is to take the enemy's country whole and intact; to shatter and destroy it is not so good. (ABS 1) **Fire is nearly guaranteed to destroy. It should not be the COA of choice.**

Chapter 12: DECEPTION PLAN
Planning; Part 2

All warfare is based on deception. (LP 18) In war, practice dissimulation, and you will succeed. (M 15) O divine art of subtlety and secrecy! Through you we learn to be invisible, through you inaudible; and hence we can hold the enemy's fate in our hands. (WPS 9)

Hence, when able to attack, we must seem unable; when using our forces, we must seem inactive; when we are near, we must make the enemy believe we are far away; when far away, we must make him believe we are near. (LP 19)

Sun Tzu says, *when we are near, we must make the enemy believe we are far away.* He uses the words, must make. He did not say we must hope, he says *must make*. Deception is both active, and passive. Moving in the low ground, or keeping below the radar may hide your actions, but what does the enemy believe instead? How are you shaping what the enemy believes? What do you want the enemy to believe? These are the considerations of a deception plan.

The Art of War is clear in the importance of deception, though it is difficult to nail down precisely when to use each suggestion. It is as much a mind-set as it is a doctrine. The key to take away is that a deception plan is necessary.

Often, the deception plan is an afterthought and never given any real consideration in the planning process. By default the deception plan becomes a reliance on secrecy. Deception in those cases is completely passive and any spur-of-the-moment actions are uncoordinated with the larger scheme. The deception plan's potential to influence events is an untapped resource for the majority of people and organizations.

See all of the action verbs in the statements below.

Hold out baits to entice the enemy. *Feign* disorder, and crush him. (LP 20) By *holding out* baits, he keeps him on the march; then with a body of picked men he *lies in wait* for him. (E 20). Thus one who is skillful at keeping the enemy on the move *maintains* deceitful appearances, according to which the enemy will act. He *sacrifices* something that the enemy may snatch at it. (E 19) By *holding out* advantages to him, he can cause the enemy to approach of his own accord; or, by *inflicting* damage, he can make it impossible for the enemy to draw near. (WPS 3)

...*Pretend* to be weak, that he may grow arrogant. (LP 22) *Simulated* disorder postulates perfect discipline, *simulated* fear postulates courage; *simulated* weakness postulates strength. (E 17) *Hiding* order beneath the cloak of disorder is simply a question of subdivision; *concealing* courage under a *show* of timidity presupposes a fund of latent energy; *masking* strength with weakness is to be effected by tactical dispositions. (E 18) In making tactical dispositions, the highest pitch you can attain is to *conceal* them; *conceal* your dispositions, and you will be safe from the prying of the subtlest spies, from the machinations of the wisest brains. (WPS 25)

He will conquer who has learnt the artifice of deviation. Such is the art of maneuvering. (M 22) Thus, to *take* a long and circuitous route, after *enticing* the enemy out of the way and though starting after him, to contrive to reach the goal before him, shows knowledge of the artifice of deviation. (M 4) Let your plans be dark and impenetrable as night, and when you move, fall like a thunderbolt. (M 19)

Appear at points which the enemy must hasten to defend; *march* swiftly to places where you are not expected.

(WPS 5) 24. *Attack* him where he is unprepared, appear where you are not expected. (LP 24) By discovering the enemy's dispositions and remaining invisible ourselves, we can keep our forces concentrated, while the enemy's must be divided. (WPS 13) We can form a single united body, while the enemy must split up into fractions. Hence there will be a whole pitted against separate parts of a whole, which means that we shall be many to the enemy's few. (WPS 14) The spot where we intend to fight must not be made known; for then the enemy will have to prepare against a possible attack at several different points; and his forces being thus distributed in many directions, the numbers we shall have to face at any given point will be proportionately few. (WPS 16)

Numerical weakness comes from having to prepare against possible attacks; numerical strength, from compelling our adversary to make these preparations against us. (WPS 18)

It is the business of a general to *be quiet* and thus ensure secrecy; upright and just, and thus maintain order. (TNS 35) By *altering* his arrangements and *changing* his plans, he keeps the enemy without definite knowledge. By *shifting* his camp and taking circuitous routes, he prevents the enemy from anticipating his purpose. (TNS 37)

Confront your soldiers with the deed itself; never let them know your design. (TNS 57) He burns his boats and breaks his cooking-pots; like a shepherd driving a flock of sheep, he drives his men this way and that, and nothing knows whither he is going. (TNS 39)

Hence he does not strive to ally himself with all and sundry, nor does he foster the power of other states. He carries out his own secret designs, keeping his antagonists in awe. Thus he is able to capture their cities and overthrow their kingdoms. (TNS 55) Reduce the hostile chiefs by inflicting damage on them; and make trouble for them, and keep them constantly engaged; hold out specious allurements, and make them rush to any given point. (VIT 10)

Having doomed spies, *doing* certain things openly for purposes of deception, and allowing our spies to know of them and report them to the enemy. (UOS 12) It is owing to his information, again, that we can cause the doomed spy to *carry* false tidings to the enemy. (UOS 23)

Chapter 13: FULL SPECTRUM DECISION MAKING
The Stubby Pencil Method

Most people do not even know that the multi-criteria decision making process (MCDMP) exists and therefore do not realize that *The Art of War* fits perfectly into this format. Please be aware that this chapter will not specifically focus on *The Art of War* though I will reinforce the connections I have already made to *The Art of War* and the MCDMP. I decided to include this chapter on the Full-Spectrum Decision Making (FSDM) process because I have touched on it in many places throughout this work and it was time to round out your understanding. Although I have modified the MCDMP into the FSDM process for easy use with a pencil and paper, the process by nature does require a significant amount of explanation, which is beyond the scope of this work. The information on the general MCDMP is widely available, but again, it is mostly in the convoluted language of academics, consultants and programmers.

My work has required practical solutions. The *Full Spectrum Decision Making Process* and the *Create the Know* program are the products of that work and here's what I've found.

Benefits that lead to Victory
By utilizing the *Pattern of Power* with the FSDM, you will be able to;

- **See what others cannot see.** Organize and Step through information in a way that allows you to see opportunities that people, who are not using this system, will miss.
- **Handle massive amounts of information.** Be able to prioritize information and focus on the most important aspects of a problem, even though there seems to be too much information to process.
- **Make the complex understandable.** Prioritization, organization, justification in a way that allows everyone to see why a decision was made.
- **Be decisive.** Confidence in a course of action comes when the scope of the problem is identified, and courses of action are designed to meet the critical needs in the best possible way. People can then decide and execute with confidence.
- **Effectively engage your imagination.** Imagination or intuition within the framework of priorities and objectives is powerful. This process does not decide for you, it gives you the opportunity to be imaginative and innovative while still solving the problem.
- **Lead and communicate with a team.** Organize and distribute the proper information at the proper times to the proper people. Then manage and lead the confidently-selected COA.

These benefits sound an awful lot like Sun Tzu's descriptions of a great general.

All men can see the tactics whereby I conquer, but what none can see is the strategy out of which victory is evolved.
(WPS 27)

There are eight steps in the FSDM process. Comparing them to the decision-making flow chart used earlier is a valid exercise and will give you a better idea of how nicely they fit together.

The ART of WAR: Organized for Decision Making

You will begin to see why I organized the *Pattern of Power* the way I did without the need to contort or degrade *The Art of War*.

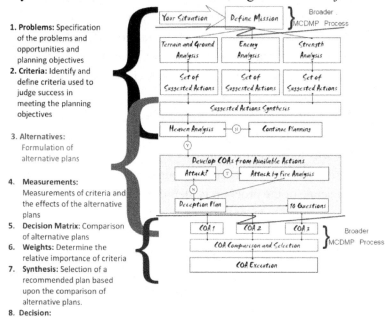

1. **Problems:** Specification of the problems and opportunities and planning objectives
2. **Criteria:** Identify and define criteria used to judge success in meeting the planning objectives
3. **Alternatives:** Formulation of alternative plans
4. **Measurements:** Measurements of criteria and the effects of the alternative plans
5. **Decision Matrix:** Comparison of alternative plans
6. **Weights:** Determine the relative importance of criteria
7. **Synthesis:** Selection of a recommended plan based upon the comparison of alternative plans.
8. **Decision:**

Multi-criteria decision making is hard work, and it offers a sometimes-unwelcome objectivity. A decision is always easier to make when we consider only one dimension of the problem and when we are the only decision maker. It was easier to look at *The Art of War* as a bunch of separate suggestions rather than a coherent philosophy.

Full-Spectrum Decision Making
FSDM is a structured approach to creating innovative solutions to complex, information-rich and multi-objective problems. The process;
- Identifies conflicts
- Distinguishes that which we know objectively from that which we do not
- Reveals the extent to which our decisions are arbitrary and based on intuition or politics
- Helps us to identify and understand conflicts and trade-offs.

The Eight Steps Expanded
In this final narrative of the *Pattern of Power*, I will expand upon the eight steps of the process. You have made some incredible leaps to become an advanced strategist and that cannot be taken away, though heed the last step of this process if you intend to actually *BE* an advanced strategist. At the same time, do not expect to master the FSDM from this summary.

Pull out a pencil and some paper and let's start.

1. **Problems:** You have to identify the scope of the problem, including where you are going, by when, and the obstacles that stand in your way. This step includes an analysis of our resources, the enemy, and the terrain you will traverse. This is a huge exercise in gathering and organizing information on all of these areas as you get your arms around your problem. As you go, you should determine which pieces of

information are facts and which are assumptions. Facts can be supported and are provable. Assumptions are drawn from a fact and are necessary for your analysis.

From the beginning, you must build a list of research items. Each and every assumption is a point of research. By recording your research items, you are certain that nothing is slipping through the cracks. It is your goal to make sure that every assumption is turned into a fact by your research. Through your relentless tracking down of facts and ferreting out of assumptions where they hide, your proposed solutions will be robust and highly stable.

On the paper in front of you, you should have tons of notes and a clearly articulated mission statement consisting of the answers to the questions, who, what, when, where and why. You have a long list of items to start researching, and notes on your friendly situation, the enemy, terrain and environment. You will have started scratching some great ideas for ways to accomplish the mission with the comfort of knowing that no matter how crazy the idea, it will be refined through the remainder of the process and evaluated with respect to your success criteria.

2. **Criteria:** Criteria are used for judging how well your alternative plans fulfill your mission. There are some measures that are nearly universally important to judging a plan such as cost, time and quality. These are characteristics of the solution that you would like to maximize or minimize. No solution to a complex problem will ever fully satisfy every criterion. Each COA will satisfy the set of criteria to different degrees. Spend time breaking down your mission so that you have a complete set of clearly defined criteria so that no significant impact goes unmeasured. Along with the Problem step, this is very hard brain work. The better you endure the details and avoid the temptation to jump right into

course of action development, the better your ultimate decision will be.
3. **Alternatives:** The fundamental purpose of multi-criteria analysis is to select the best course of action from among a set of alternatives. Your mission and subsequent analysis are used to guide the construction of courses of action. The COA answers the question, "how are you going to accomplish your mission and best fulfill your criteria?" A COA is otherwise known as a plan.

Brainstorming and engaging a flow of your own ideas are methods for creating alternative courses of action. The process is characterized by imagination, innovation and freedom. Force yourself to construct at least three separate and distinct COAs.

Each COA will have a main effort and several supporting efforts. The main effort will be those initiatives that are most important to defeating the enemy and accomplishing the mission. The supporting efforts do just that; they support the main effort. The accomplishment of the supporting efforts is essential to the success of the main effort.

Your deception plan is a supporting effort.

This format is one way the framework forces ingenuity and innovation. You are allowed to be *wild* because you know that through the later parts of the model the impacts of *ingenious* twists in your plan will be evaluated. Make each COA as detailed as necessary to describe clearly how you plan to use your resources to achieve your mission. Be sure, as you start to mold your separate COAs, that you continue to record research items such as assumptions, resources needed, and potential impacts on your criteria.

The COA impacts are studied, estimated and evaluated routinely in the planning process and are explicitly incorporated into the latter steps of the decision support framework. The COA formulation step is a critical part of the multi-criteria analysis. In fact, many of the strongest selling points for multi-criteria analysis have to do with the structured thought process it imposes on decision making. Without alternatives from which to choose, there is no need for a structured process. When you have only one idea, the decision is much easier.

At the end of this step you will have at least three COAs with distinct main efforts. You will have described your actions in each COA from the beginning to end and estimated the needed resources for each plan. I know, lots more paper.

4. **Measurements:** Your work becomes easier from here on out if you have done a thorough job in your planning up to this point. Measure the effects that each COA has on your list of criteria. Place an actual value per criteria per COA. COA 1 will cost $"A," take "B" days and net us "C" acres. COA 2 will cost $"X," take "Y" days and net us "Z" acres. Fill in the ABC's and XYZ's.

5. **Decision Matrix:** The decision matrix is a method of compiling all of our critical information into a table format. It provides a clear way for structuring and presenting the problem. Criteria are listed down the left side and the COAs are listed across the top of the table. The measurements for each criteria are noted in the corresponding cells.

You should see by now how your judgment on each COA is tied directly to our criteria, which were derived from our mission. You have stayed focused on your mission, while inviting innovation to take part in the process.

6. **Weights:** Determine the relative importance of criteria. Some criteria are more important for us to fulfill than others.

By applying a higher numerical value to our higher weighted criteria, they will have a relatively higher impact in determining which COA is selected. This step allows us to better see how each COA fulfills our planning objectives.

7. **Synthesis:** The synthesis combines all of the data you have gathered, along with the measurements and weights of your criteria. This provides you the ability to compare the COAs side by side to get a good idea of how well each will fulfill your mission. Before you launch off into selecting a COA because it *scores* the best, you must do a sensitivity analysis. The purpose of this exercise is to ensure that the results of your synthesis do not rely too heavily on any assumption or a group of assumptions. The less sensitive your solution is to things not known for certain, the more robust it will be.

   If you have not done the process properly you are fooling yourself as to the sturdiness and accuracy of your plan. This may be the time to take a breath, get a cup of coffee and come back to your solutions with a fresh eye.

8. **Decision: Do what the overwhelming majority will never do. Make a decision and execute it.**

CONCLUSION to The Pattern of Power

1. Proper planning provides the insights to effective action.
2. Action is the distinguishing feature of the achiever's life.
3. Perseverance, consistent action in the face of opposition, is the secret to great achievement.

I have introduced four titled concepts that fit together in overlapping ways; *The Pattern of Power, The Art of War*, Multi-Criteria Decision Making Process and the Full-Spectrum Decision Making Process. The FSDM, as it is presented here encompasses and incorporates the important parts of the other processes.

The Art of War is the vaunted 2,500 year-old philosophy written by Sun Tzu. My contention is that it is brilliant, but disorganized for the 21st century strategist.

The Pattern of Power is the reorganization of *The Art of War* into a functional, usable tool for you.

The Multi-Criteria Decision Making Process (MCDMP) allows for timely, confident and competent decisions for complex problems involving massive amounts of information and conflicting priorities.

The Full-Spectrum Decision Making Process (FCDM) is a customized and simplified MCDMP. The FCDM incorporates *The Pattern of Power* as a major portion of the process. This merger of philosophy and process was not possible prior to the reorganization of the information in *The Art of War*.

Therefore, the FCDM along with *The Pattern of Power* allow you to see through Sun Tzu's eyes like no other work. Use it properly and you will be able to;
- See what others cannot see
- Handle massive amounts of information
- Make the complex understandable
- Be decisive
- Effectively engage their imagination
- Lead and communicate with a team

                Go and achieve Supreme Excellence!

THE ART OF WAR

The Oldest Military Treatise in the World. Translated from the Chinese By Lionel Giles, M.A. (1910)

I. LAYING PLANS

1. Sun Tzu said: *The Art of War* is of vital importance to the State.
2. It is a matter of life and death, a road either to safety or to ruin. Hence it is a subject of inquiry which can on no account be neglected.
3. *The Art of War*, then, is governed by five constant factors, to be taken into account in one's deliberations, when seeking to determine the conditions obtaining in the field.
4. These are: (1) The Moral Law; (2) Heaven; (3) Earth; (4) The Commander; (5) Method and discipline.
5,6. The Moral Law causes the people to be in complete accord with their ruler, so that they will follow him regardless of their lives, undismayed by any danger.
7. Heaven signifies night and day, cold and heat, times and seasons.
8. Earth comprises distances, great and small; danger and security; open ground and narrow passes; the chances of life and death.
9. The Commander stands for the virtues of wisdom, sincerely, benevolence, courage and strictness.
10. By method and discipline are to be understood the marshaling of the army in its proper subdivisions, the graduations of rank among the officers, the maintenance of roads

by which supplies may reach the army, and the control of military expenditure.

11. These five heads should be familiar to every general: he who knows them will be victorious; he who knows them not will fail.

12. Therefore, in your deliberations, when seeking to determine the military conditions, let them be made the basis of a comparison, in this wise:--

13. (1) Which of the two sovereigns is imbued with the Moral law? (2) Which of the two generals has most ability? (3) With whom lie the advantages derived from Heaven and Earth? (4) On which side is discipline most rigorously enforced? (5) Which army is stronger? (6) On which side are officers and men more highly trained? (7) In which army is there the greater constancy both in reward and punishment?

14. By means of these seven considerations I can forecast victory or defeat.

15. The general that hearkens to my counsel and acts upon it, will conquer: let such a one be retained in command! The general that hearkens not to my counsel nor acts upon it, will suffer defeat:--let such a one be dismissed!

16. While heading the profit of my counsel, avail yourself also of any helpful circumstances over and beyond the ordinary rules.

17. According as circumstances are favorable, one should modify one's plans.

18. All warfare is based on deception.

19. Hence, when able to attack, we must seem unable; when using our forces, we must seem inactive; when we are near, we must make the enemy believe we are far away; when far away, we must make him believe we are near.

20. Hold out baits to entice the enemy. Feign disorder, and crush him.

21. If he is secure at all points, be prepared for him. If he is in superior strength, evade him.

22. If your opponent is of choleric temper, seek to irritate him. Pretend to be weak, that he may grow arrogant.

23. If he is taking his ease, give him no rest. If his forces are united, separate them.

24. Attack him where he is unprepared, appear where you are not expected.

25. These military devices, leading to victory, must not be divulged beforehand.

26. Now the general who wins a battle makes many calculations in his temple ere the battle is fought. The general who loses a battle makes but few calculations beforehand. Thus do many calculations lead to victory, and few calculations to defeat: how much more no calculation at all! It is by attention to this point that I can foresee who is likely to win or lose.

II. WAGING WAR

1. Sun Tzu said: In the operations of war, where there are in the field a thousand swift chariots, as many heavy chariots, and a hundred thousand mail-clad soldiers, with provisions enough to carry them a thousand li, the expenditure at home and at the front, including entertainment of guests, small items such as glue and paint, and sums spent on chariots and armor, will reach the total of a thousand ounces of silver per day. Such is the cost of raising an army of 100,000 men.

2. When you engage in actual fighting, if victory is long in coming, then men's weapons will grow dull and their ardor will be damped. If you lay siege to a town, you will exhaust your strength.

3. Again, if the campaign is protracted, the resources of the State will not be equal to the strain.

4. Now, when your weapons are dulled, your ardor damped, your strength exhausted and your treasure spent, other chieftains will

spring up to take advantage of your extremity. Then no man, however wise, will be able to avert the consequences that must ensue.

5. Thus, though we have heard of stupid haste in war, cleverness has never been seen associated with long delays.

6. There is no instance of a country having benefited from prolonged warfare.

7. It is only one who is thoroughly acquainted with the evils of war that can thoroughly understand the profitable way of carrying it on.

8. The skillful soldier does not raise a second levy, neither are his supply-wagons loaded more than twice.

9. Bring war material with you from home, but forage on the enemy. Thus the army will have food enough for its needs.

10. Poverty of the State exchequer causes an army to be maintained by contributions from a distance. Contributing to maintain an army at a distance causes the people to be impoverished.

11. On the other hand, the proximity of an army causes prices to go up; and high prices cause the people's substance to be drained away.

12. When their substance is drained away, the peasantry will be afflicted by heavy exactions.

13,14. With this loss of substance and exhaustion of strength, the homes of the people will be stripped bare, and three-tenths of their income will be dissipated; while government expenses for broken chariots, worn-out horses, breast-plates and helmets, bows and arrows, spears and shields, protective mantles, draught-oxen and heavy wagons, will amount to four-tenths of its total revenue.

15. Hence a wise general makes a point of foraging on the enemy. One cartload of the enemy's provisions is equivalent to twenty of one's own, and likewise a single picul of his provender is equivalent to twenty from one's own store.

16. Now in order to kill the enemy, our men must be roused to anger; that there may be advantage from defeating the enemy, they must have their rewards.

17. Therefore in chariot fighting, when ten or more chariots have been taken, those should be rewarded who took the first. Our own flags should be substituted for those of the enemy, and the chariots mingled and used in conjunction with ours. The captured soldiers should be kindly treated and kept.

18. This is called, using the conquered foe to augment one's own strength.

19. In war, then, let your great object be victory, not lengthy campaigns.

20. Thus it may be known that the leader of armies is the arbiter of the people's fate, the man on whom it depends whether the nation shall be in peace or in peril.

III. ATTACK BY STRATAGEM

1. Sun Tzu said: In the practical art of war, the best thing of all is to take the enemy's country whole and intact; to shatter and destroy it is not so good. So, too, it is better to recapture an army entire than to destroy it, to capture a regiment, a detachment or a company entire than to destroy them.

2. Hence to fight and conquer in all your battles is not supreme excellence; supreme excellence consists in breaking the enemy's resistance without fighting.

3. Thus the highest form of generalship is to balk the enemy's plans; the next best is to prevent the junction of the enemy's forces; the next in order is to attack the enemy's army in the field; and the worst policy of all is to besiege walled cities.

4. The rule is, not to besiege walled cities if it can possibly be avoided. The preparation of mantlets, movable shelters, and various implements of war, will take up three whole months; and

the piling up of mounds over against the walls will take three months more.

5. The general, unable to control his irritation, will launch his men to the assault like swarming ants, with the result that one-third of his men are slain, while the town still remains untaken. Such are the disastrous effects of a siege.

6. Therefore the skillful leader subdues the enemy's troops without any fighting; he captures their cities without laying siege to them; he overthrows their kingdom without lengthy operations in the field.

7. With his forces intact he will dispute the mastery of the Empire, and thus, without losing a man, his triumph will be complete. This is the method of attacking by stratagem.

8. It is the rule in war, if our forces are ten to the enemy's one, to surround him; if five to one, to attack him; if twice as numerous, to divide our army into two.

9. If equally matched, we can offer battle; if slightly inferior in numbers, we can avoid the enemy; if quite unequal in every way, we can flee from him.

10. Hence, though an obstinate fight may be made by a small force, in the end it must be captured by the larger force.

11. Now the general is the bulwark of the State; if the bulwark is complete at all points; the State will be strong; if the bulwark is defective, the State will be weak.

12. There are three ways in which a ruler can bring misfortune upon his army:--

13. (1) By commanding the army to advance or to retreat, being ignorant of the fact that it cannot obey. This is called hobbling the army.

14. (2) By attempting to govern an army in the same way as he administers a kingdom, being ignorant of the conditions which obtain in an army. This causes restlessness in the soldier's minds.

15. (3) By employing the officers of his army without discrimination, through ignorance of the military principle of

adaptation to circumstances. This shakes the confidence of the soldiers.

16. But when the army is restless and distrustful, trouble is sure to come from the other feudal princes. This is simply bringing anarchy into the army, and flinging victory away.

17. Thus we may know that there are five essentials for victory: (1) He will win who knows when to fight and when not to fight. (2) He will win who knows how to handle both superior and inferior forces. (3) He will win whose army is animated by the same spirit throughout all its ranks. (4) He will win who, prepared himself, waits to take the enemy unprepared. (5) He will win who has military capacity and is not interfered with by the sovereign.

18. Hence the saying: If you know the enemy and know yourself, you need not fear the result of a hundred battles. If you know yourself but not the enemy, for every victory gained you will also suffer a defeat. If you know neither the enemy nor yourself, you will succumb in every battle.

IV. TACTICAL DISPOSITIONS

1. Sun Tzu said: The good fighters of old first put themselves beyond the possibility of defeat, and then waited for an opportunity of defeating the enemy.

2. To secure ourselves against defeat lies in our own hands, but the opportunity of defeating the enemy is provided by the enemy himself.

3. Thus the good fighter is able to secure himself against defeat, but cannot make certain of defeating the enemy.

4. Hence the saying: One may know how to conquer without being able to do it.

5. Security against defeat implies defensive tactics; ability to defeat the enemy means taking the offensive.

6. Standing on the defensive indicates insufficient strength; attacking, a superabundance of strength.

7. The general who is skilled in defense hides in the most secret recesses of the earth; he who is skilled in attack flashes forth from the topmost heights of heaven. Thus on the one hand we have ability to protect ourselves; on the other, a victory that is complete.

8. To see victory only when it is within the ken of the common herd is not the acme of excellence.

9. Neither is it the acme of excellence if you fight and conquer and the whole Empire says, "Well done!"

10. To lift an autumn hair is no sign of great strength; to see the sun and moon is no sign of sharp sight; to hear the noise of thunder is no sign of a quick ear.

11. What the ancients called a clever fighter is one who not only wins, but excels in winning with ease.

12. Hence his victories bring him neither reputation for wisdom nor credit for courage.

13. He wins his battles by making no mistakes. Making no mistakes is what establishes the certainty of victory, for it means conquering an enemy that is already defeated.

14. Hence the skillful fighter puts himself into a position which makes defeat impossible, and does not miss the moment for defeating the enemy.

15. Thus it is that in war the victorious strategist only seeks battle after the victory has been won, whereas he who is destined to defeat first fights and afterwards looks for victory.

16. The consummate leader cultivates the moral law, and strictly adheres to method and discipline; thus it is in his power to control success.

17. In respect of military method, we have, firstly, Measurement; secondly, Estimation of quantity; thirdly, Calculation; fourthly, Balancing of chances; fifthly, Victory.

18. Measurement owes its existence to Earth; Estimation of quantity to Measurement; Calculation to Estimation of quantity; Balancing of chances to Calculation; and Victory to Balancing of chances.
19. A victorious army opposed to a routed one, is as a pound's weight placed in the scale against a single grain.
20. The onrush of a conquering force is like the bursting of pent-up waters into a chasm a thousand fathoms deep.

V. ENERGY

1. Sun Tzu said: The control of a large force is the same principle as the control of a few men: it is merely a question of dividing up their numbers.
2. Fighting with a large army under your command is nowise different from fighting with a small one: it is merely a question of instituting signs and signals.
3. To ensure that your whole host may withstand the brunt of the enemy's attack and remain unshaken – this is effected by maneuvers direct and indirect.
4. That the impact of your army may be like a grindstone dashed against an egg – this is effected by the science of weak points and strong.
5. In all fighting, the direct method may be used for joining battle, but indirect methods will be needed in order to secure victory.
6. Indirect tactics, efficiently applied, are inexhaustible as Heaven and Earth, unending as the flow of rivers and streams; like the sun and moon, they end but to begin anew; like the four seasons, they pass away to return once more.
7. There are not more than five musical notes, yet the combinations of these five give rise to more melodies than can ever be heard.

8. There are not more than five primary colors (blue, yellow, red, white, and black), yet in combination they produce more hues than can ever been seen.

9. There are not more than five cardinal tastes (sour, acrid, salt, sweet, bitter), yet combinations of them yield more flavors than can ever be tasted.

10. In battle, there are not more than two methods of attack--the direct and the indirect; yet these two in combination give rise to an endless series of maneuvers.

11. The direct and the indirect lead on to each other in turn. It is like moving in a circle--you never come to an end. Who can exhaust the possibilities of their combination?

12. The onset of troops is like the rush of a torrent which will even roll stones along in its course.

13. The quality of decision is like the well-timed swoop of a falcon which enables it to strike and destroy its victim.

14. Therefore the good fighter will be terrible in his onset, and prompt in his decision.

15. Energy may be likened to the bending of a crossbow; decision, to the releasing of a trigger.

16. Amid the turmoil and tumult of battle, there may be seeming disorder and yet no real disorder at all; amid confusion and chaos, your array may be without head or tail, yet it will be proof against defeat.

17. Simulated disorder postulates perfect discipline, simulated fear postulates courage; simulated weakness postulates strength.

18. Hiding order beneath the cloak of disorder is simply a question of subdivision; concealing courage under a show of timidity presupposes a fund of latent energy; masking strength with weakness is to be effected by tactical dispositions.

19. Thus one who is skillful at keeping the enemy on the move maintains deceitful appearances, according to which the enemy will act. He sacrifices something, that the enemy may snatch at it.

20. By holding out baits, he keeps him on the march; then with a body of picked men he lies in wait for him.

21. The clever combatant looks to the effect of combined energy, and does not require too much from individuals. Hence his ability to pick out the right men and utilize combined energy.

22. When he utilizes combined energy, his fighting men become as it were like unto rolling logs or stones. For it is the nature of a log or stone to remain motionless on level ground, and to move when on a slope; if four-cornered, to come to a standstill, but if round-shaped, to go rolling down.

23. Thus the energy developed by good fighting men is as the momentum of a round stone rolled down a mountain thousands of feet in height. So much on the subject of energy.

VI. WEAK POINTS AND STRONG

1. Sun Tzu said: Whoever is first in the field and awaits the coming of the enemy, will be fresh for the fight; whoever is second in the field and has to hasten to battle will arrive exhausted.

2. Therefore the clever combatant imposes his will on the enemy, but does not allow the enemy's will to be imposed on him.

3. By holding out advantages to him, he can cause the enemy to approach of his own accord; or, by inflicting damage, he can make it impossible for the enemy to draw near.

4. If the enemy is taking his ease, he can harass him; if well supplied with food, he can starve him out; if quietly encamped, he can force him to move.

5. Appear at points which the enemy must hasten to defend; march swiftly to places where you are not expected.

6. An army may march great distances without distress, if it marches through country where the enemy is not.

7. You can be sure of succeeding in your attacks if you only attack places which are undefended. You can ensure the safety of your defense if you only hold positions that cannot be attacked.
8. Hence that general is skillful in attack whose opponent does not know what to defend; and he is skillful in defense whose opponent does not know what to attack.
9. O divine art of subtlety and secrecy! Through you we learn to be invisible, through you inaudible; and hence we can hold the enemy's fate in our hands.
10. You may advance and be absolutely irresistible, if you make for the enemy's weak points; you may retire and be safe from pursuit if your movements are more rapid than those of the enemy.
11. If we wish to fight, the enemy can be forced to an engagement even though he be sheltered behind a high rampart and a deep ditch. All we need do is attack some other place that he will be obliged to relieve.
12. If we do not wish to fight, we can prevent the enemy from engaging us even though the lines of our encampment be merely traced out on the ground. All we need do is to throw something odd and unaccountable in his way.
13. By discovering the enemy's dispositions and remaining invisible ourselves, we can keep our forces concentrated, while the enemy's must be divided.
14. We can form a single united body, while the enemy must split up into fractions. Hence there will be a whole pitted against separate parts of a whole, which means that we shall be many to the enemy's few.
15. And if we are able thus to attack an inferior force with a superior one, our opponents will be in dire straits.
16. The spot where we intend to fight must not be made known; for then the enemy will have to prepare against a possible attack at several different points; and his forces being thus distributed

in many directions, the numbers we shall have to face at any given point will be proportionately few.

17. For should the enemy strengthen his van, he will weaken his rear; should he strengthen his rear, he will weaken his van; should he strengthen his left, he will weaken his right; should he strengthen his right, he will weaken his left. If he sends reinforcements everywhere, he will everywhere be weak.

18. Numerical weakness comes from having to prepare against possible attacks; numerical strength, from compelling our adversary to make these preparations against us.

19. Knowing the place and the time of the coming battle, we may concentrate from the greatest distances in order to fight.

20. But if neither time nor place be known, then the left wing will be impotent to succour the right, the right equally impotent to succour the left, the van unable to relieve the rear, or the rear to support the van. How much more so if the furthest portions of the army are anything under a hundred LI apart, and even the nearest are separated by several LI!

21. Though according to my estimate the soldiers of Yueh exceed our own in number, that shall advantage them nothing in the matter of victory. I say then that victory can be achieved.

22. Though the enemy be stronger in numbers, we may prevent him from fighting. Scheme so as to discover his plans and the likelihood of their success.

23. Rouse him, and learn the principle of his activity or inactivity. Force him to reveal himself, so as to find out his vulnerable spots.

24. Carefully compare the opposing army with your own, so that you may know where strength is superabundant and where it is deficient.

25. In making tactical dispositions, the highest pitch you can attain is to conceal them; conceal your dispositions, and you will be safe from the prying of the subtlest spies, from the machinations of the wisest brains.

26. How victory may be produced for them out of the enemy's own tactics--that is what the multitude cannot comprehend.
27. All men can see the tactics whereby I conquer, but what none can see is the strategy out of which victory is evolved.
28. Do not repeat the tactics which have gained you one victory, but let your methods be regulated by the infinite variety of circumstances.
29. Military tactics are like unto water; for water in its natural course runs away from high places and hastens downwards.
30. So in war, the way is to avoid what is strong and to strike at what is weak.
31. Water shapes its course according to the nature of the ground over which it flows; the soldier works out his victory in relation to the foe whom he is facing.
32. Therefore, just as water retains no constant shape, so in warfare there are no constant conditions.
33. He who can modify his tactics in relation to his opponent and thereby succeed in winning, may be called a heaven-born captain.
34. The five elements (water, fire, wood, metal, earth) are not always equally predominant; the four seasons make way for each other in turn. There are short days and long; the moon has its periods of waning and waxing.

VII. MANEUVERING

1. Sun Tzu said: In war, the general receives his commands from the sovereign.
2. Having collected an army and concentrated his forces, he must blend and harmonize the different elements thereof before pitching his camp.
3. After that, comes tactical maneuvering, than which there is nothing more difficult. The difficulty of tactical maneuvering

consists in turning the devious into the direct, and misfortune into gain.

4. Thus, to take a long and circuitous route, after enticing the enemy out of the way, and though starting after him, to contrive to reach the goal before him, shows knowledge of the artifice of DEVIATION.

5. Maneuvering with an army is advantageous; with an undisciplined multitude, most dangerous.

6. If you set a fully equipped army in march in order to snatch an advantage, the chances are that you will be too late. On the other hand, to detach a flying column for the purpose involves the sacrifice of its baggage and stores.

7. Thus, if you order your men to roll up their buff-coats, and make forced marches without halting day or night, covering double the usual distance at a stretch, doing a hundred LI in order to wrest an advantage, the leaders of all your three divisions will fall into the hands of the enemy.

8. The stronger men will be in front, the jaded ones will fall behind, and on this plan only one-tenth of your army will reach its destination.

9. If you march fifty LI in order to outmaneuver the enemy, you will lose the leader of your first division, and only half your force will reach the goal.

10. If you march thirty LI with the same object, two-thirds of your army will arrive.

11. We may take it then that an army without its baggage-train is lost; without provisions it is lost; without bases of supply it is lost.

12. We cannot enter into alliances until we are acquainted with the designs of our neighbors.

13. We are not fit to lead an army on the march unless we are familiar with the face of the country--its mountains and forests, its pitfalls and precipices, its marshes and swamps.

14. We shall be unable to turn natural advantage to account unless we make use of local guides.
15. In war, practice dissimulation, and you will succeed.
16. Whether to concentrate or to divide your troops, must be decided by circumstances.
17. Let your rapidity be that of the wind, your compactness that of the forest.
18. In raiding and plundering be like fire, is immovability like a mountain.
19. Let your plans be dark and impenetrable as night, and when you move, fall like a thunderbolt.
20. When you plunder a countryside, let the spoil be divided amongst your men; when you capture new territory, cut it up into allotments for the benefit of the soldiery.
21. Ponder and deliberate before you make a move.
22. He will conquer who has learnt the artifice of deviation. Such is the art of maneuvering.
23. The Book of Army Management says: On the field of battle, the spoken word does not carry far enough: hence the institution of gongs and drums. Nor can ordinary objects be seen clearly enough: hence the institution of banners and flags.
24. Gongs and drums, banners and flags, are means whereby the ears and eyes of the host may be focused on one particular point.
25. The host thus forming a single united body, is it impossible either for the brave to advance alone, or for the cowardly to retreat alone. This is the art of handling large masses of men.
26. In night-fighting, then, make much use of signal-fires and drums, and in fighting by day, of flags and banners, as a means of influencing the ears and eyes of your army.
27. A whole army may be robbed of its spirit; a commander-in-chief may be robbed of his presence of mind.
28. Now a soldier's spirit is keenest in the morning; by noonday it has begun to flag; and in the evening, his mind is bent only on returning to camp.

29. A clever general, therefore, avoids an army when its spirit is keen, but attacks it when it is sluggish and inclined to return. This is the art of studying moods.

30. Disciplined and calm, to await the appearance of disorder and hubbub amongst the enemy:--this is the art of retaining self-possession.

31. To be near the goal while the enemy is still far from it, to wait at ease while the enemy is toiling and struggling, to be well-fed while the enemy is famished:--this is the art of husbanding one's strength.

32. To refrain from intercepting an enemy whose banners are in perfect order, to refrain from attacking an army drawn up in calm and confident array:--this is the art of studying circumstances.

33. It is a military axiom not to advance uphill against the enemy, nor to oppose him when he comes downhill.

34. Do not pursue an enemy who simulates flight; do not attack soldiers whose temper is keen.

35. Do not swallow bait offered by the enemy. Do not interfere with an army that is returning home.

36. When you surround an army, leave an outlet free. Do not press a desperate foe too hard.

37. Such is *The Art of War*fare.

VIII. VARIATION IN TACTICS

1. Sun Tzu said: In war, the general receives his commands from the sovereign, collects his army and concentrates his forces

2. When in difficult country, do not encamp. In country where high roads intersect, join hands with your allies. Do not linger in dangerously isolated positions. In hemmed-in situations, you must resort to stratagem. In desperate position, you must fight.

3. There are roads which must not be followed, armies which must be not attacked, towns which must be besieged, positions

which must not be contested, commands of the sovereign which must not be obeyed.

4. The general who thoroughly understands the advantages that accompany variation of tactics knows how to handle his troops.

5. The general who does not understand these, may be well acquainted with the configuration of the country, yet he will not be able to turn his knowledge to practical account.

6. So, the student of war who is unversed in *The Art of War* of varying his plans, even though he be acquainted with the Five Advantages, will fail to make the best use of his men.

7. Hence in the wise leader's plans, considerations of advantage and of disadvantage will be blended together.

8. If our expectation of advantage be tempered in this way, we may succeed in accomplishing the essential part of our schemes.

9. If, on the other hand, in the midst of difficulties we are always ready to seize an advantage, we may extricate ourselves from misfortune.

10. Reduce the hostile chiefs by inflicting damage on them; and make trouble for them, and keep them constantly engaged; hold out specious allurements, and make them rush to any given point.

11. *The Art of War* teaches us to rely not on the likelihood of the enemy's not coming, but on our own readiness to receive him; not on the chance of his not attacking, but rather on the fact that we have made our position unassailable.

12. There are five dangerous faults which may affect a general: (1) Recklessness, which leads to destruction; (2) cowardice, which leads to capture; (3) a hasty temper, which can be provoked by insults; (4) a delicacy of honor which is sensitive to shame; (5) over-solicitude for his men, which exposes him to worry and trouble.

13. These are the five besetting sins of a general, ruinous to the conduct of war.

14. When an army is overthrown and its leader slain, the cause will surely be found among these five dangerous faults. Let them be a subject of meditation.

IX. THE ARMY ON THE MARCH

1. Sun Tzu said: We come now to the question of encamping the army, and observing signs of the enemy. Pass quickly over mountains, and keep in the neighborhood of valleys.
2. Camp in high places, facing the sun. Do not climb heights in order to fight. So much for mountain warfare.
3. After crossing a river, you should get far away from it.
4. When an invading force crosses a river in its onward march, do not advance to meet it in mid-stream. It will be best to let half the army get across, and then deliver your attack.
5. If you are anxious to fight, you should not go to meet the invader near a river which he has to cross.
6. Moor your craft higher up than the enemy, and facing the sun. Do not move up-stream to meet the enemy. So much for river warfare.
7. In crossing salt-marshes, your sole concern should be to get over them quickly, without any delay.
8. If forced to fight in a salt-marsh, you should have water and grass near you, and get your back to a clump of trees. So much for operations in salt-marches.
9. In dry, level country, take up an easily accessible position with rising ground to your right and on your rear, so that the danger may be in front, and safety lie behind. So much for campaigning in flat country.
10. These are the four useful branches of military knowledge which enabled the Yellow Emperor to vanquish four several sovereigns.

11. All armies prefer high ground to low and sunny places to dark.
12. If you are careful of your men, and camp on hard ground, the army will be free from disease of every kind, and this will spell victory.
13. When you come to a hill or a bank, occupy the sunny side, with the slope on your right rear. Thus you will at once act for the benefit of your soldiers and utilize the natural advantages of the ground.
14. When, in consequence of heavy rains up-country, a river which you wish to ford is swollen and flecked with foam, you must wait until it subsides.
15. Country in which there are precipitous cliffs with torrents running between, deep natural hollows, confined places, tangled thickets, quagmires and crevasses, should be left with all possible speed and not approached.
16. While we keep away from such places, we should get the enemy to approach them; while we face them, we should let the enemy have them on his rear.
17. If in the neighborhood of your camp there should be any hilly country, ponds surrounded by aquatic grass, hollow basins filled with reeds, or woods with thick undergrowth, they must be carefully routed out and searched; for these are places where men in ambush or insidious spies are likely to be lurking.
18. When the enemy is close at hand and remains quiet, he is relying on the natural strength of his position.
19. When he keeps aloof and tries to provoke a battle, he is anxious for the other side to advance.
20. If his place of encampment is easy of access, he is tendering a bait.
21. Movement amongst the trees of a forest shows that the enemy is advancing. The appearance of a number of screens in the midst of thick grass means that the enemy wants to make us suspicious.

22. The rising of birds in their flight is the sign of an ambuscade. Startled beasts indicate that a sudden attack is coming.

23. When there is dust rising in a high column, it is the sign of chariots advancing; when the dust is low, but spread over a wide area, it betokens the approach of infantry. When it branches out in different directions, it shows that parties have been sent to collect firewood. A few clouds of dust moving to and fro signify that the army is encamping.

24. Humble words and increased preparations are signs that the enemy is about to advance. Violent language and driving forward as if to the attack are signs that he will retreat.

25. When the light chariots come out first and take up a position on the wings, it is a sign that the enemy is forming for battle.

26. Peace proposals unaccompanied by a sworn covenant indicate a plot.

27. When there is much running about and the soldiers fall into rank, it means that the critical moment has come.

28. When some are seen advancing and some retreating, it is a lure.

29. When the soldiers stand leaning on their spears, they are faint from want of food.

30. If those who are sent to draw water begin by drinking themselves, the army is suffering from thirst.

31. If the enemy sees an advantage to be gained and makes no effort to secure it, the soldiers are exhausted.

32. If birds gather on any spot, it is unoccupied. Clamour by night betokens nervousness.

33. If there is disturbance in the camp, the general's authority is weak. If the banners and flags are shifted about, sedition is afoot. If the officers are angry, it means that the men are weary.

34. When an army feeds its horses with grain and kills its cattle for food, and when the men do not hang their cooking-pots over the camp-fires, showing that they will not return to their tents, you may know that they are determined to fight to the death.

35. The sight of men whispering together in small knots or speaking in subdued tones points to disaffection amongst the rank and file.
36. Too frequent rewards signify that the enemy is at the end of his resources; too many punishments betray a condition of dire distress.
37. To begin by bluster, but afterwards to take fright at the enemy's numbers, shows a supreme lack of intelligence.
38. When envoys are sent with compliments in their mouths, it is a sign that the enemy wishes for a truce.
39. If the enemy's troops march up angrily and remain facing ours for a long time without either joining battle or taking themselves off again, the situation is one that demands great vigilance and circumspection.
40. If our troops are no more in number than the enemy, that is amply sufficient; it only means that no direct attack can be made. What we can do is simply to concentrate all our available strength, keep a close watch on the enemy, and obtain reinforcements.
41. He who exercises no forethought but makes light of his opponents is sure to be captured by them.
42. If soldiers are punished before they have grown attached to you, they will not prove submissive; and, unless submissive, then will be practically useless. If, when the soldiers have become attached to you, punishments are not enforced, they will still be unless.
43. Therefore soldiers must be treated in the first instance with humanity, but kept under control by means of iron discipline. This is a certain road to victory.
44. If in training soldiers commands are habitually enforced, the army will be well-disciplined; if not, its discipline will be bad.
45. If a general shows confidence in his men but always insists on his orders being obeyed, the gain will be mutual.

X. TERRAIN

1. Sun Tzu said: We may distinguish six kinds of terrain, to wit: (1) Accessible ground; (2) entangling ground; (3) temporizing ground; (4) narrow passes; (5) precipitous heights; (6) positions at a great distance from the enemy.
2. Ground which can be freely traversed by both sides is called accessible.
3. With regard to ground of this nature, be before the enemy in occupying the raised and sunny spots, and carefully guard your line of supplies. Then you will be able to fight with advantage.
4. Ground which can be abandoned but is hard to re-occupy is called entangling.
5. From a position of this sort, if the enemy is unprepared, you may sally forth and defeat him. But if the enemy is prepared for your coming, and you fail to defeat him, then, return being impossible, disaster will ensue.
6. When the position is such that neither side will gain by making the first move, it is called temporizing ground.
7. In a position of this sort, even though the enemy should offer us an attractive bait, it will be advisable not to stir forth, but rather to retreat, thus enticing the enemy in his turn; then, when part of his army has come out, we may deliver our attack with advantage.
8. With regard to narrow passes, if you can occupy them first, let them be strongly garrisoned and await the advent of the enemy.
9. Should the army forestall you in occupying a pass, do not go after him if the pass is fully garrisoned, but only if it is weakly garrisoned.
10. With regard to precipitous heights, if you are beforehand with your adversary, you should occupy the raised and sunny spots, and there wait for him to come up.
11. If the enemy has occupied them before you, do not follow him, but retreat and try to entice him away.

12. If you are situated at a great distance from the enemy, and the strength of the two armies is equal, it is not easy to provoke a battle, and fighting will be to your disadvantage.

13. These six are the principles connected with Earth. The general who has attained a responsible post must be careful to study them.

14. Now an army is exposed to six several calamities, not arising from natural causes, but from faults for which the general is responsible. These are: (1) Flight; (2) insubordination; (3) collapse; (4) ruin; (5) disorganization; (6) rout.

15. Other conditions being equal, if one force is hurled against another ten times its size, the result will be the flight of the former.

16. When the common soldiers are too strong and their officers too weak, the result is insubordination. When the officers are too strong and the common soldiers too weak, the result is collapse.

17. When the higher officers are angry and insubordinate, and on meeting the enemy give battle on their own account from a feeling of resentment, before the commander-in-chief can tell whether or no he is in a position to fight, the result is ruin.

18. When the general is weak and without authority; when his orders are not clear and distinct; when there are no fixes duties assigned to officers and men, and the ranks are formed in a slovenly haphazard manner, the result is utter disorganization.

19. When a general, unable to estimate the enemy's strength, allows an inferior force to engage a larger one, or hurls a weak detachment against a powerful one, and neglects to place picked soldiers in the front rank, the result must be rout.

20. These are six ways of courting defeat, which must be carefully noted by the general who has attained a responsible post.

21. The natural formation of the country is the soldier's best ally; but a power of estimating the adversary, of controlling the forces

of victory, and of shrewdly calculating difficulties, dangers and distances, constitutes the test of a great general.

22. He who knows these things, and in fighting puts his knowledge into practice, will win his battles. He who knows them not, nor practices them, will surely be defeated.

23. If fighting is sure to result in victory, then you must fight, even though the ruler forbid it; if fighting will not result in victory, then you must not fight even at the ruler's bidding.

24. The general who advances without coveting fame and retreats without fearing disgrace, whose only thought is to protect his country and do good service for his sovereign, is the jewel of the kingdom.

25. Regard your soldiers as your children, and they will follow you into the deepest valleys; look upon them as your own beloved sons, and they will stand by you even unto death.

26. If, however, you are indulgent, but unable to make your authority felt; kind-hearted, but unable to enforce your commands; and incapable, moreover, of quelling disorder: then your soldiers must be likened to spoilt children; they are useless for any practical purpose.

27. If we know that our own men are in a condition to attack, but are unaware that the enemy is not open to attack, we have gone only halfway towards victory.

28. If we know that the enemy is open to attack, but are unaware that our own men are not in a condition to attack, we have gone only halfway towards victory.

29. If we know that the enemy is open to attack, and also know that our men are in a condition to attack, but are unaware that the nature of the ground makes fighting impracticable, we have still gone only halfway towards victory.

30. Hence the experienced soldier, once in motion, is never bewildered; once he has broken camp, he is never at a loss.

31. Hence the saying: If you know the enemy and know yourself, your victory will not stand in doubt; if you know Heaven and know Earth, you may make your victory complete.

XI. THE NINE SITUATIONS

1. Sun Tzu said: *The Art of War* recognizes nine varieties of ground: (1) Dispersive ground; (2) facile ground; (3) contentious ground; (4) open ground; (5) ground of intersecting highways; (6) serious ground; (7) difficult ground; (8) hemmed-in ground; (9) desperate ground.
2. When a chieftain is fighting in his own territory, it is dispersive ground.
3. When he has penetrated into hostile territory, but to no great distance, it is facile ground.
4. Ground the possession of which imports great advantage to either side, is contentious ground.
5. Ground on which each side has liberty of movement is open ground.
6. Ground which forms the key to three contiguous states, so that he who occupies it first has most of the Empire at his command, is a ground of intersecting highways.
7. When an army has penetrated into the heart of a hostile country, leaving a number of fortified cities in its rear, it is serious ground.
8. Mountain forests, rugged steeps, marshes and fens--all country that is hard to traverse: this is difficult ground.
9. Ground which is reached through narrow gorges, and from which we can only retire by tortuous paths, so that a small number of the enemy would suffice to crush a large body of our men: this is hemmed in ground.
10. Ground on which we can only be saved from destruction by fighting without delay, is desperate ground.

11. On dispersive ground, therefore, fight not. On facile ground, halt not. On contentious ground, attack not.
12. On open ground, do not try to block the enemy's way. On the ground of intersecting highways, join hands with your allies.
13. On serious ground, gather in plunder. In difficult ground, keep steadily on the march.
14. On hemmed-in ground, resort to stratagem. On desperate ground, fight.
15. Those who were called skillful leaders of old knew how to drive a wedge between the enemy's front and rear; to prevent cooperation between his large and small divisions; to hinder the good troops from rescuing the bad, the officers from rallying their men.
16. When the enemy's men were united, they managed to keep them in disorder.
17. When it was to their advantage, they made a forward move; when otherwise, they stopped still.
18. If asked how to cope with a great host of the enemy in orderly array and on the point of marching to the attack, I should say: "Begin by seizing something which your opponent holds dear; then he will be amenable to your will."
19. Rapidity is the essence of war: take advantage of the enemy's unreadiness, make your way by unexpected routes, and attack unguarded spots.
20. The following are the principles to be observed by an invading force: The further you penetrate into a country, the greater will be the solidarity of your troops, and thus the defenders will not prevail against you.
21. Make forays in fertile country in order to supply your army with food.
22. Carefully study the well-being of your men, and do not overtax them. Concentrate your energy and hoard your strength. Keep your army continually on the move, and devise unfathomable plans.

23. Throw your soldiers into positions whence there is no escape, and they will prefer death to flight. If they will face death, there is nothing they may not achieve. Officers and men alike will put forth their uttermost strength.

24. Soldiers when in desperate straits lose the sense of fear. If there is no place of refuge, they will stand firm. If they are in hostile country, they will show a stubborn front. If there is no help for it, they will fight hard.

25. Thus, without waiting to be marshaled, the soldiers will be constantly on the qui vive; without waiting to be asked, they will do your will; without restrictions, they will be faithful; without giving orders, they can be trusted.

26. Prohibit the taking of omens, and do away with superstitious doubts. Then, until death itself comes, no calamity need be feared.

27. If our soldiers are not overburdened with money, it is not because they have a distaste for riches; if their lives are not unduly long, it is not because they are disinclined to longevity.

28. On the day they are ordered out to battle, your soldiers may weep, those sitting up bedewing their garments, and those lying down letting the tears run down their cheeks. But let them once be brought to bay, and they will display the courage of a Chu or a Kuei.

29. The skillful tactician may be likened to the shuai-jan. Now the shuai-jan is a snake that is found in the ChUng mountains. Strike at its head, and you will be attacked by its tail; strike at its tail, and you will be attacked by its head; strike at its middle, and you will be attacked by head and tail both.

30. Asked if an army can be made to imitate the shuai-jan, I should answer, Yes. For the men of Wu and the men of Yueh are enemies; yet if they are crossing a river in the same boat and are caught by a storm, they will come to each other's assistance just as the left hand helps the right.

31. Hence it is not enough to put one's trust in the tethering of horses, and the burying of chariot wheels in the ground

32. The principle on which to manage an army is to set up one standard of courage which all must reach.

33. How to make the best of both strong and weak--that is a question involving the proper use of ground.

34. Thus the skillful general conducts his army just as though he were leading a single man, willy-nilly, by the hand.

35. It is the business of a general to be quiet and thus ensure secrecy; upright and just, and thus maintain order.

36. He must be able to mystify his officers and men by false reports and appearances, and thus keep them in total ignorance.

37. By altering his arrangements and changing his plans, he keeps the enemy without definite knowledge. By shifting his camp and taking circuitous routes, he prevents the enemy from anticipating his purpose.

38. At the critical moment, the leader of an army acts like one who has climbed up a height and then kicks away the ladder behind him. He carries his men deep into hostile territory before he shows his hand.

39. He burns his boats and breaks his cooking-pots; like a shepherd driving a flock of sheep, he drives his men this way and that, and nothing knows whither he is going.

40. To muster his host and bring it into danger:--this may be termed the business of the general.

41. The different measures suited to the nine varieties of ground; the expediency of aggressive or defensive tactics; and the fundamental laws of human nature: these are things that must most certainly be studied.

42. When invading hostile territory, the general principle is, that penetrating deeply brings cohesion; penetrating but a short way means dispersion.

43. When you leave your own country behind, and take your army across neighborhood territory, you find yourself on critical

ground. When there are means of communication on all four sides, the ground is one of intersecting highways.

44. When you penetrate deeply into a country, it is serious ground. When you penetrate but a little way, it is facile ground.

45. When you have the enemy's strongholds on your rear, and narrow passes in front, it is hemmed-in ground. When there is no place of refuge at all, it is desperate ground.

46. Therefore, on dispersive ground, I would inspire my men with unity of purpose. On facile ground, I would see that there is close connection between all parts of my army.

47. On contentious ground, I would hurry up my rear.

48. On open ground, I would keep a vigilant eye on my defenses. On ground of intersecting highways, I would consolidate my alliances.

49. On serious ground, I would try to ensure a continuous stream of supplies. On difficult ground, I would keep pushing on along the road.

50. On hemmed-in ground, I would block any way of retreat. On desperate ground, I would proclaim to my soldiers the hopelessness of saving their lives.

51. For it is the soldier's disposition to offer an obstinate resistance when surrounded, to fight hard when he cannot help himself, and to obey promptly when he has fallen into danger.

52. We cannot enter into alliance with neighboring princes until we are acquainted with their designs. We are not fit to lead an army on the march unless we are familiar with the face of the country--its mountains and forests, its pitfalls and precipices, its marshes and swamps. We shall be unable to turn natural advantages to account unless we make use of local guides.

53. To be ignored of any one of the following four or five principles does not befit a warlike prince.

54. When a warlike prince attacks a powerful state, his generalship shows itself in preventing the concentration of the

enemy's forces. He overawes his opponents, and their allies are prevented from joining against him.

55. Hence he does not strive to ally himself with all and sundry, nor does he foster the power of other states. He carries out his own secret designs, keeping his antagonists in awe. Thus he is able to capture their cities and overthrow their kingdoms.

56. Bestow rewards without regard to rule, issue orders without regard to previous arrangements; and you will be able to handle a whole army as though you had to do with but a single man.

57. Confront your soldiers with the deed itself; never let them know your design. When the outlook is bright, bring it before their eyes; but tell them nothing when the situation is gloomy.

58. Place your army in deadly peril, and it will survive; plunge it into desperate straits, and it will come off in safety.

59. For it is precisely when a force has fallen into harm's way that is capable of striking a blow for victory.

60. Success in warfare is gained by carefully accommodating ourselves to the enemy's purpose.

61. By persistently hanging on the enemy's flank, we shall succeed in the long run in killing the commander-in-chief.

62. This is called ability to accomplish a thing by sheer cunning.

63. On the day that you take up your command, block the frontier passes, destroy the official tallies, and stop the passage of all emissaries.

64. Be stern in the council-chamber, so that you may control the situation.

65. If the enemy leaves a door open, you must rush in.

66. Forestall your opponent by seizing what he holds dear, and subtly contrive to time his arrival on the ground.

67. Walk in the path defined by rule, and accommodate yourself to the enemy until you can fight a decisive battle.

68. At first, then, exhibit the coyness of a maiden, until the enemy gives you an opening; afterwards emulate the rapidity of

a running hare, and it will be too late for the enemy to oppose you.

XII. THE ATTACK BY FIRE

1. Sun Tzu said: There are five ways of attacking with fire. The first is to burn soldiers in their camp; the second is to burn stores; the third is to burn baggage trains; the fourth is to burn arsenals and magazines; the fifth is to hurl dropping fire amongst the enemy.
2. In order to carry out an attack, we must have means available. The material for raising fire should always be kept in readiness.
3. There is a proper season for making attacks with fire, and special days for starting a conflagration.
4. The proper season is when the weather is very dry; the special days are those when the moon is in the constellations of the Sieve, the Wall, the Wing or the Cross-bar; for these four are all days of rising wind.
5. In attacking with fire, one should be prepared to meet five possible developments:
6. (1) When fire breaks out inside to enemy's camp, respond at once with an attack from without.
7. (2) If there is an outbreak of fire, but the enemy's soldiers remain quiet, bide your time and do not attack.
8. (3) When the force of the flames has reached its height, follow it up with an attack, if that is practicable; if not, stay where you are.
9. (4) If it is possible to make an assault with fire from without, do not wait for it to break out within, but deliver your attack at a favorable moment.
10. (5) When you start a fire, be to windward of it. Do not attack from the leeward.

11. A wind that rises in the daytime lasts long, but a night breeze soon falls.

12. In every army, the five developments connected with fire must be known, the movements of the stars calculated, and a watch kept for the proper days.

13. Hence those who use fire as an aid to the attack show intelligence; those who use water as an aid to the attack gain an accession of strength.

14. By means of water, an enemy may be intercepted, but not robbed of all his belongings.

15. Unhappy is the fate of one who tries to win his battles and succeed in his attacks without cultivating the spirit of enterprise; for the result is waste of time and general stagnation.

16. Hence the saying: The enlightened ruler lays his plans well ahead; the good general cultivates his resources.

17. Move not unless you see an advantage; use not your troops unless there is something to be gained; fight not unless the position is critical.

18. No ruler should put troops into the field merely to gratify his own spleen; no general should fight a battle simply out of pique.

19. If it is to your advantage, make a forward move; if not, stay where you are.

20. Anger may in time change to gladness; vexation may be succeeded by content.

21. But a kingdom that has once been destroyed can never come again into being; nor can the dead ever be brought back to life.

22. Hence the enlightened ruler is heedful, and the good general full of caution. This is the way to keep a country at peace and an army intact.

XIII. THE USE OF SPIES

1. Sun Tzu said: Raising a host of a hundred thousand men and marching them great distances entails heavy loss on the people and a drain on the resources of the State. The daily expenditure will amount to a thousand ounces of silver. There will be commotion at home and abroad, and men will drop down exhausted on the highways. As many as seven hundred thousand families will be impeded in their labor.

2. Hostile armies may face each other for years, striving for the victory which is decided in a single day. This being so, to remain in ignorance of the enemy's condition simply because one grudges the outlay of a hundred ounces of silver in honors and emoluments, is the height of inhumanity.

3. One who acts thus is no leader of men, no present help to his sovereign, no master of victory.

4. Thus, what enables the wise sovereign and the good general to strike and conquer, and achieve things beyond the reach of ordinary men, is foreknowledge.

5. Now this foreknowledge cannot be elicited from spirits; it cannot be obtained inductively from experience, nor by any deductive calculation.

6. Knowledge of the enemy's dispositions can only be obtained from other men.

7. Hence the use of spies, of whom there are five classes: (1) Local spies; (2) inward spies; (3) converted spies; (4) doomed spies; (5) surviving spies.

8. When these five kinds of spy are all at work, none can discover the secret system. This is called "divine manipulation of the threads." It is the sovereign's most precious faculty.

9. Having local spies means employing the services of the inhabitants of a district.

10. Having inward spies, making use of officials of the enemy.

11. Having converted spies, getting hold of the enemy's spies and using them for our own purposes.

12. Having doomed spies, doing certain things openly for purposes of deception, and allowing our spies to know of them and report them to the enemy.

13. Surviving spies, finally, are those who bring back news from the enemy's camp.

14. Hence it is that which none in the whole army are more intimate relations to be maintained than with spies. None should be more liberally rewarded. In no other business should greater secrecy be preserved.

15. Spies cannot be usefully employed without a certain intuitive sagacity.

16. They cannot be properly managed without benevolence and straightforwardness.

17. Without subtle ingenuity of mind, one cannot make certain of the truth of their reports.

18. Be subtle! be subtle! and use your spies for every kind of business.

19. If a secret piece of news is divulged by a spy before the time is ripe, he must be put to death together with the man to whom the secret was told.

20. Whether the object be to crush an army, to storm a city, or to assassinate an individual, it is always necessary to begin by finding out the names of the attendants, the aides-de-camp, and door-keepers and sentries of the general in command. Our spies must be commissioned to ascertain these.

21. The enemy's spies who have come to spy on us must be sought out, tempted with bribes, led away and comfortably housed. Thus they will become converted spies and available for our service.

22. It is through the information brought by the converted spy that we are able to acquire and employ local and inward spies.

23. It is owing to his information, again, that we can cause the doomed spy to carry false tidings to the enemy.

24. Lastly, it is by his information that the surviving spy can be used on appointed occasions.

25. The end and aim of spying in all its five varieties is knowledge of the enemy; and this knowledge can only be derived, in the first instance, from the converted spy. Hence it is essential that the converted spy be treated with the utmost liberality.

26. Of old, the rise of the Yin dynasty was due to I Chih who had served under the Hsia. Likewise, the rise of the Chou dynasty was due to Lu Ya who had served under the Yin.

27. Hence it is only the enlightened ruler and the wise general who will use the highest intelligence of the army for purposes of spying and thereby they achieve great results. Spies are a most important element in water, because on them depends an army's ability to move.

© Patrol Leader Press and Brace E. Barber
www.NoExcuseLeadership.com
b@mkoi.us

Made in the USA
Middletown, DE
23 November 2020